Evolutionary Ecology of Parasites

Evolutionary Ecology of Parasites

From individuals to communities

Robert Poulin

Department of Zoology
University of Otago
Dunedin
New Zealand

CHAPMAN & HALL
London · Weinheim · New York · Tokyo · Melbourne · Madras

Published by Chapman & Hall, 2–6 Boundary Row, London SE1 8HN, UK

Chapman & Hall, 2–6 Boundary Row, London SE1 8HN, UK

Chapman & Hall GmbH, Pappelallee 3, 69469 Weinheim, Germany

Chapman & Hall USA, 115 Fifth Avenue, New York, NY 10003, USA

Chapman & Hall Japan, ITP-Japan, Kyowa Building, 3F, 2-2-1 Hirakawacho, Chiyoda-ku, Tokyo 102, Japan

Chapman & Hall Australia, 102 Dodds Street, South Melbourne, Victoria 3205, Australia

Chapman & Hall India, R. Seshadri, 32 Second Main Road, CIT East, Madras 600 035, India

First edition 1998

Typeset in 10/12pt Palatino by Saxon Graphics Ltd, Derby
Printed in Great Britain

ISBN 0 412 80560 X (HB) 0 412 793709 (PB)

A catalogue record for this book is available from the British Library

∞ Printed on permanent acid-free text paper, manufactured in accordance with ANSI/NISO Z39.48-1992 and ANSI-NISO Z39.48-1984 (Permanence of Paper).

Contents

Preface

Nothing can be more rewarding to a student of parasitism than the sudden and recent popularity of parasites among ecologists and evolutionary biologists. After years on a dusty shelf under the label 'gross and disgusting', parasites have recently been the subject of articles in popular scientific journals such as *BioScience, New Scientist* and *Scientific American.* It is somewhat frustrating, however, that there are hardly any textbooks covering the many recent developments in the evolutionary and ecological study of parasites. Most texts of evolutionary biology and ecology just do not mention parasites, while many parasitology texts provide only a phylum-by-phylum description of parasite life cycles and fail to discuss the exciting aspects of their biology. The lack of a good, recent text exploring parasite evolution and ecology prompted me to write this book, in the hope that readers will then see parasites as fascinating creatures rather than boring worms. After I got started, Combes' (1995) book came out and filled part of the gap in the literature. There was still a need for a concise book, more focused on parasites than hosts and aimed at more advanced students of parasitology; this is what I intended this book to be.

A few warnings are necessary at this stage. Firstly, I have tried to present the big picture instead of becoming lost in details. This book deals with the ultimate or evolutionary level of biology, rather than the proximate or mechanistic level. I wanted to answer questions relating to 'why' parasites do certain things rather than 'how' they do these things. This approach has in the past brought me criticism from 'classically' trained, hands-on parasitologists, and will no doubt earn me some more. In particular, many people are quick to point out exceptions to general patterns in a valiant effort to refute those patterns. I welcome such comments, but instead I choose to view the exceptions as special cases requiring further explanation. In any event, most ideas presented here are backed up with empirical evidence; I have been selective in my choice of examples to illustrate the conceptual framework, and I do not present exhaustive lists of case studies.

Secondly, this book comprises the themes and ideas that survived the elimination process when going through my mind's filter. The content is therefore biased toward what I think are today's most important research areas and theoretical concepts. Other authors would undoubtedly come up with different chapter topics and examples. I have also opted to cite recent studies and reviews rather than older material. Readers keen to trace back the history of ideas to their very beginning should look up those references.

This book is intended as a reference for practising parasitologists, and hopefully will be useful to other evolutionary ecologists. I have assumed a basic knowledge and understanding of ecology, evolution and parasite biology. The book could thus only be used as a textbook for an advanced undergraduate course or, more probably, for a graduate course in parasitology. My hope is that it will stimulate both young and established researchers to use parasites as model organisms in evolutionary ecology studies.

Many people have encouraged and helped me, either intellectually or emotionally, during the writing of this book. They deserve much more credit than I can give them in these few lines: without their advice and support there would probably be no book. Alison Dunn, Dave Marcogliese, Janice Moore, Serge Morand, Adrian Paterson, Klaus Rohde, Mike Sukhdeo, Frédéric Thomas and Dave Wharton have all made valuable comments on one or more chapters. I have not always followed their recommendations, and any unorthodox interpretation in the following pages is strictly my own. Andrew Stammer formerly of Chapman & Hall was enthusiastic about the project and played a major role in getting the ball rolling in the early stages. Finally, I thank Diane, whose supportive presence continuously erased the doubts arising in my mind about the book, and Eric and Alexandre, our own little parasites who did an excellent job taking my mind off the writing when a break was needed.

Introduction 1

Ecology is the scientific study of interactions between organisms of the same or different species, and between organisms and their non-living environment. One of the main goals of ecologists is to explain the abundance and distribution of organisms over space and time. The scope of ecology includes all sorts of interactions, from the most intimate, permanent associations to the briefest of encounters. Although parasitism qualifies as the sort of interaction of interest to ecologists, it has become the focus for another branch of science, parasitology, which uses a multi-disciplinary approach to investigate host-parasite interactions. Because of the intimate and intricate nature of the association between host and parasite, a broadly trained parasitologist using techniques ranging from molecular biology to ecology seemed the most appropriate investigator. However, parasites have thus been ignored by ecologists for a long time. Sections on parasitism have only recently begun to appear in ecology textbooks (for instance, Begon *et al.*, 1996 and earlier editions), and these bear mainly on the ecological impacts of parasites on free-living organisms. Studies on the population or community ecology of parasites themselves are practically absent from the ecology literature and are almost exclusively restricted to parasitology journals.

Similarly, parasite evolution has until recently been studied by parasitologists rather than by evolutionary biologists. As with ecological studies, evolutionary investigations of host-parasite interactions undertaken by non-parasitologists are recent and still very few (Poulin, 1995a). While the quality of studies on the ecology or evolution of parasitism performed by parasitologists is not in doubt, it is unfortunate that for many years there have been few exchanges of ideas between parasitology and either ecology or evolutionary biology. Researchers in these disciplines attend different meetings and read different journals. This has lead to philosophical differences between them, some of which are important. For instance, parasitologists have long known that parasites can affect host population dynamics, but ecologists took some time to realize this. Also, until recently and in part because of the influence of medical science on parasitology, most parasitologists believed that evolution led to a decrease in parasite virulence, whereas modern evolutionary theory would have predicted a greater range of outcomes (Ewald, 1994). These disagreements could have been avoided had there been a better integration of ecology, evolutionary studies and knowledge of parasite biology by students of parasitism.

My purpose in this book is to present an evolutionary ecologist's view of the biology of parasites. I want to re-examine aspects of the biology of parasites using an approach compatible with current theory in evolutionary biology and ecology. Many studies of parasite ecology or evolution published in the past 40 or 50 years were thorough descriptive investigations but failed to test the general hypotheses put forward by ecologists and evolutionary biologists; here I will try to rectify this by emphasizing the link with theory. In this book I approach parasites as an evolutionary ecologist would approach any other group of organisms, while recognizing the special attributes of parasites. The book focuses on parasites themselves rather than on their interaction with hosts. Instead, hosts are seen as a key part of the parasites' environment and as a major source of selective pressures. The influence of parasites on host biology has been dealt with extensively in recent reviews; it will be covered here only as it relates to the ecology or evolution of parasites.

1.1 THE EVOLUTIONARY ECOLOGY APPROACH

Organisms interact with one another and with the non-living environment on an ecological time scale, measured in days or years. These interactions, however, are the product of natural selection acting over evolutionary time, over thousands and millions of years, to produce organisms well suited to their environment. Evolutionary ecology is the study of the selective pressures imposed by the environment and the evolutionary responses to these pressures. Natural selection has shaped not only the traits of individual organisms, but also the properties of populations and species assemblages. The subject matter of evolutionary ecology therefore includes topics such as the trade-off between the size and number of offspring produced by individual animals; the proportion of males and females in animal populations; and the composition of animal communities. All these phenomena can be studied on a human time scale but to understand the differences observed among organisms, one must consider the forces and constraints that have acted during their evolutionary history.

The study of evolutionary responses is not always as straightforward as that of phenomena occurring on shorter time scales. A major goal of science is to demonstrate causality: it can be inferred that an event causes a response if the response always follows the event in an experimental situation. For example, exsheathment of many nematode larvae and hatching of many cestode eggs always follow their exposure to the conditions encountered in their host's gut in *in vitro* experiments, therefore it can safely be inferred that these conditions cause exsheathment or hatching. In evolutionary ecology, the manipulation of variables in controlled experiments is usually impossible. Instead, we must rely heavily on comparisons between species that have been exposed to different selective pressures. If species under a given selective regime have consistently evolved a certain combination of traits, these 'natural' experiments can be used to draw conclusions about the effects of certain factors over evolutionary time. Obviously, similarities between species can be the result of inheritance of traits from a common ancestor as much as the product of independent lines of convergent evolution. A careful distinction must be made between phylogenetic influences and the true action of natural selection (Brooks and McLennan, 1991; Harvey and Pagel, 1991). In the absence of other evidence, comparisons across taxa can help to identify true adaptations, defined here as genetically determined traits that have spread or are spreading through a population because they confer greater fitness on their bearers.

Although only applied recently to parasitological problems, the comparative approach can shed much light on parasite evolution. This approach can do more than identify relationships between species traits. It can also be used to test evolutionary hypotheses, even though it does not follow the classical experimental approach consisting of the manipulation of independent variables in controlled conditions (see Brandon, 1994). Different parasite lineages leading to extant species can be viewed as different evolutionary experiments, in which the ancestor represents the initial experimental conditions and the current phenotype represents the experimental outcome. Comparing lineages evolving under different selective pressures (e.g. in different types of hosts) is like comparing the responses of subjects exposed to different experimental conditions, or their responses to the manipulation of selected variables. In this context, controlling for phylogenetic influences corresponds to avoiding pseudoreplication.

Proper comparative analyses are powerful tools for hypothesis testing in evolutionary ecology. They are not, however, a panacea for the study of adaptation. Used in isolation from other kinds of evidence, comparative studies provide limited insights into evolutionary mechanisms and the causal links between biological traits (Doughty, 1996). On the other hand, the comparative approach is the most useful to identify general patterns that can guide further research. Although the results of controlled experiments or field observations are used as tests of theory where possible, much of the evidence presented in this book relies on the explanation of variability among species using a comparative approach.

This book is not an elementary treatise of evolutionary ecology. The reader who wants a more general overview of the theory and mathematical models at the core of modern evolutionary ecology can read any of several recent texts on the subject (e.g. Cockburn, 1991; Bulmer, 1994; Pianka, 1994). This book applies many ideas from evolutionary ecology specifically to parasites, and aims to foster the use of evolutionary thinking in the study of parasite ecology. Some of the questions that will be addressed include: Why do some parasites have more complex life cycles than others? Why are some parasites more host specific than others? Why are some parasites much more fecund than others? Why are some parasites much more virulent than others? Why are some parasites more highly aggregated among their hosts than others? Why are some parasite communities richer than others? These questions have been addressed before by parasitologists, but usually not in an evolutionary context nor with appropriate comparative methods.

1.2 SCOPE AND OVERVIEW

Because of its vague definition, the term parasite has been applied to a wide range of plant and animal taxa. The most widely accepted definition of a parasite is that it is an organism living in or on another organism, the host – feeding on it, showing some degree of structural adaptation to it, and causing it some harm. Interpretations of this definition vary among authors. Price (1980) included phytophagous insects as parasites, but excluded blood-sucking flies. Barnard (1990) included behavioural parasites, such as many birds which are not physiologically dependent on their host but exploit them in other ways, for example by stealing food from the host. Combes (1995) even included strands of DNA among parasitic entities. It is therefore necessary to specify the taxonomic scope of this book, which will focus exclusively on protozoan and metazoan parasites of animals. These include several diverse taxa of parasites (Table 1.1) that have had several independent evolutionary origins. Because helminths have been the subject of the majority of relevant studies, they will provide most examples. The general biology and life cycles of these parasites are described in detail in any basic parasitology text (e.g. Noble *et al.*, 1989; Schmidt and Roberts, 1989; Cox, 1993), and it will be assumed that the reader is at least superficially familiar with them. Some of the patterns discussed here and some of the conclusions they suggest may also apply to other groups of parasites, but these are beyond the scope of this book.

The evolutionary ecology of parasites can be studied at several hierarchical levels. The smallest unit of study in ecology is the individual

Table 1.1 Diversity of some of the major taxa of metazoan parasites of animals. Estimates of species numbers were obtained from various sources and are meant to be realistic minimum numbers

Parasite taxon	Number of species	Definitive host[a]	Life cycle[b]	Habitat[c]
Phylum Platyhelminthes				
Class Trematoda	>15000	Endo, V	C	M, FW, T
Class Monogenea	>20000	Ecto, V	S	M, FW
Class Cestoidea	>5000	Endo, V	C	M, FW, T
Phylum Acanthocephala	>1200	Endo, V	C	M, FW, T
Phylum Nematomorpha	>350	Endo, I	S&C	FW, T
Phylum Nematoda	>10000	Endo, I&V	S&C	M, FW, T
Phylum Mollusca				
Class Bivalvia	>600	Ecto, V	S	FW
Phylum Annelida				
Class Hirudinea	>400	Ecto, V	S	M, FW
Phylum Arthropoda				
Subphylum Chelicerata				
Class Arachnida	>30000	Ecto, I&V	S & C	M, FW, T
Subphylum Crustacea				
Class Branchiura	>150	Ecto, V	S	M, FW
Class Copepoda	>3800	Ecto, I&V	S	M, FW
Class Cirripedia				
Subclass Ascothoracida	>100	Endo, I	S	M
Subclass Rhizocephala	>260	Ecto, I	S	M
Class Malacostraca				
Order Isopoda	>600	Ecto, I&V	S	M
Order Amphipoda	>250	Ecto, I&V	S	M
Subphylum Uniramia				
Class Insecta				
Order Mallophaga	>3000	Ecto, V	S	T
Order Anoplura	>3000	Ecto, V	S	T
Order Siphonaptera	>2000	Ecto, V	S	T

[a] Parasites are classified as either endoparasitic (Endo) or ectoparasitic (Ecto) on either invertebrate (I) or vertebrate (V) definitive hosts.
[b] Life cycles are simple (S) if a single host is required and complex (C) if two or more hosts are required for the completion of the cycle.
[c] The habitat of parasites and their hosts can be marine (M), freshwater (FW) or terrestrial (T).

organism, but ecologists also deal with populations of individuals of the same species, and with communities made up of several populations of different species. This book first examines how ecological traits of individual parasites have evolved, and then considers population and community characteristics. Chapter 2 presents a discussion of how organisms that made a transition to parasitism from a free-living ancestral lifestyle have undergone changes in their biology, and also considers how historical events and selective pressures have shaped complex life cycles, and how these life cycles in turn have influenced the ecology of the parasites adopting them. Chapter 3 explores the reasons why some parasites have evolved the ability to exploit a wide range of hosts whereas others are restricted to a single host species. For organisms often thought to be small, degenerate egg-production machines, parasites also show a tremendous range of life-history traits. Chapter 4 discusses how much of this variation is explained by selective pressures from the host or the physical environment, and how much is due to phylogenetic constraints. The final characteristic of individual parasites discussed here is their ability to harm or manipulate the host. Far from evolving to become benign commensals, parasites can be selected to become highly virulent exploiters of host resources, or they can evolve the ability to control the physiology and behaviour of their host for their own benefit. Chapter 5 explores the conditions under which host exploitation strategies can evolve toward these extremes.

One of the easiest-described properties of animal populations is their distribution in space. Parasite populations are typically aggregated among their host populations, but the degree of aggregation varies greatly over time and among populations and species. The opening chapter on parasite population ecology will examine the causes of aggregation, and some of its potential evolutionary consequences (Chapter 6). Parasite individuals in a population are not distributed evenly in space, and their numbers also fluctuate in time. The basic models of parasite population dynamics are reviewed in Chapter 7, along with a discussion of how evolution may have shaped various population processes. In nature, any parasite population is likely to coexist with populations of other parasite species. The transition to parasite community ecology will be made by examining how populations interact and how parasites have responded to interspecific competition (Chapter 8). Parasites of different species occurring in the same host individual form a community, which itself is only a small subset of the larger community comprising all parasite species found in the host population. In turn, this is a subset of the list of parasite species known from the populations of that host species. At all these levels of organization, assemblages of parasite species may be structured or they may be random, i.e. they may be predictable sets of species taken from the pool of available species, or an assemblage formed by chance events. This and other issues are addressed at all levels of parasite community organization (Chapters 9 and 10).

The cover illustration reveals the scope of the book at a glance. This book is about the biology of individual parasites, such as their transmission pattern and life cycle; the biology of parasite populations, including their distribution among hosts; and the biology of parasite communities, with emphasis on structure and richness. All these themes are linked to one another and set within an evolutionary or phylogenetic framework. The evolutionary ecology of parasites is a young discipline and many questions remain unanswered. Throughout the text, areas in which further research is required are highlighted. Hopefully, these suggestions will lead to more investigations and a narrowing of the gap between parasite evolutionary ecology and the evolutionary ecology of free-living organisms. In Chapter 11, general guidelines are presented for future studies, and the global importance of evolutionary studies of parasites is discussed.

Origins of parasitism and complex life cycles 2

Parasites have obviously evolved from free-living ancestors – there first had to be animals around for parasites to exploit. Parasitism has originated independently in several animal taxa, and has sometimes arisen more than once in a given taxon. The origins of parasites typically go back several million years, as indicated by the available fossil evidence (Conway Morris, 1981). Since these early days parasites have diversified greatly, and modern species now display a wide variety of life cycles and adaptations to their parasitic existence. This chapter presents scenarios for the early origins and evolution of parasites and the subsequent complexity of their life cycles. Other aspects of the life history of parasites, such as the evolution of body size and reproductive output, will be covered in Chapter 4.

2.1 TRANSITIONS TO PARASITISM

Two organisms of widely different sizes may come into contact for some time without the small one exploiting or harming the large one, and without the large one eating the small one. This is probably common in nature. If these encounters are frequent in a given pair of species, an opportunity exists for a more permanent association to develop. However, opportunity is only one of the requirements for the establishment of an intimate interspecific association. The parasite-to-be must possess some pre-adaptations for survival, feeding and reproduction on the host, and its reproductive success as a parasite must be greater than its success as a free-living animal. Without pre-adaptations the parasite cannot begin to exploit the host, and without fitness benefits host exploitation will not be favoured by natural selection.

The need for pre-adaptation of parasite precursors is an old idea (Rothschild and Clay, 1952). Specialization for a parasitic existence may be gradual, but the initial stages of host exploitation must be possible at the onset. Selection would not gradually shape feeding and attachment organs from one generation to the next if the parasite could not derive benefits from the host, i.e. fitness gains must precede specialization. Thus the animals that can become parasites are the ones capable of remaining in or on the host for certain periods of time, during which they can feed on the host and achieve greater fitness than their conspecifics which do not exploit hosts. The mouth parts and other structures necessary to feed on dead animals, or to burrow into plant stems, or to suck the sap of plants, are all examples of pre-adaptation to parasitism on animals. Some gastropod molluscs (Bouchet and Perrine, 1996) and ostracod crustaceans (Stepien and Brusca, 1985), not usually known to be parasitic, have been documented feeding at night on fish resting near the sea floor. Crabs are usually free-living but larval and adult crabs can enter the gill cavity of fish caught in traps on the sea floor, and feed on their gill filaments (Williams and Bunkley-Williams, 1994). Such opportunistic foragers, using a feeding apparatus designed for a different purpose, could eventually evolve into specialized parasites forming more permanent and intimate associations with their fish hosts. Cymothoid isopods, for instance, are obligate ectoparasites of fish evolved from ancestors that were only facultative fish parasites (Brusca, 1981). The closest relatives of the Cymothoidae, e.g. Aegidae and Corallanidae, are still free-living predators of invertebrates capable of an occasional blood meal on a fish.

Many small invertebrates attach to the external surfaces of larger animals for a limited time to disperse into new areas. This phenomenon, called phoresy, is seen as a common step toward a more permanent, parasitic association. For instance, both mites and nematodes include phoretic and parasitic species. In these taxa parasitic species may have evolved from phoretic ancestors (Anderson, 1984; Athias-Binche and Morand, 1993). In particular, the dauer larvae of nematodes possess the necessary characteristics to use other animals as transport hosts, being resistant to a range of conditions including accidental passage through an animal's gut. Phoresis may be more beneficial in terrestrial than in aquatic environments; water currents facilitate dispersal in the latter. Not surprisingly, despite the rich fauna of free-living nematodes in aquatic habitats, nematodes parasitic on animals probably all originated in terrestrial habitats from phoretic ancestors, and the relatively few nematodes parasitic on aquatic hosts have terrestrial ancestors (Anderson, 1984, 1996). Whatever its exact nature, some form of pre-adaptation must have been present in the precursors of parasites to allow the association to get started; a strict physiological dependence of the parasite on the host and efficient methods of host-to-host transmission would have evolved subsequently.

How often has parasitism arisen as a mode of existence in the history of life on Earth? It is not possible to answer this question. A fully resolved animal phylogeny would be necessary just to start guessing, but an accurate answer is not possible because it is not known how many parasitic lineages have become extinct without leaving a trace. It is clear, though, that transitions from a free-living existence to parasitism have been very common. Among nematodes, parasitism of animals has evolved on many independent occasions (Anderson, 1984; Clarck, 1994). Digeneans, monogeneans and cestodes are a monophyletic group presumably descended from a single evolutionary transition to parasitism by a common ancestor; however, associations with hosts have had multiple independent origins in turbellarians, another group of parasitic flatworms (Rohde, 1994a). In crustaceans, several taxa include both parasitic and free-living species, and parasitism has arisen on several different occasions in copepods (Poulin, 1995b), isopods (Poulin, 1995c) and amphipods (Poulin and Hamilton, 1995). It is probably safe to say that living metazoan parasites of animals are the product of at least 50 separate evolutionary transitions from a free-living existence to one of obligate parasitism.

2.2 SPECIALIZATION OF PARASITES

Discussions of parasite evolution often revolve around the idea that parasites are evolutionarily retrogressive or degenerate. This perception comes from comparisons made between parasites and their hosts: obviously, worms are morphologically less complex than their vertebrate hosts, whatever definition of complexity is used. This superficial and erroneous comparison leads to the conclusion that parasites have lost sense organs and other structures during their evolution until they became the simple life forms we observe today. Such a process has been called sacculinization (after a parasite, of course, the crustacean *Sacculina*) and is still associated with parasites in the contemporary literature. As Combes (1995) points out, some functions such as digestion and locomotion are often left entirely to the host, and in these situations the loss of structures or organs without active roles makes sense. The economy of energy and resources thus achieved may have allowed other structures to develop in response to selection. The apparent loss of complexity, however, is deceptive. Studies using DNA reassociation techniques have indicated that the genome of parasites is often larger and more complex that that of their free-living relatives, not what would be expected if parasites evolve to become simpler. For example, the parasitic nematodes *Ascaris* and *Trichinella* have much larger genomes than those of free-living nematodes, and the cestode *Hymenolepis* has a genome twice as complex as that of the free-living flatworm *Dugesia* (Searcy and MacInnis, 1970). Ironically, the genome of the cestode *Bothriocephalus gregarius* is twice the size of that of its vertebrate host (Verneau *et al.*, 1991). The morphological complexity of parasites does not always reflect their genomic complexity.

Perhaps the complexity is partitioned among the different stages in the life cycle, so that a complex genome codes for an ontogenetic succession of relatively simple forms. Viewing the whole life cycle, and not just the adult form, as the unit of selection gives a more accurate perspective on parasite complexity.

With the aid of modern electron microscopic techniques, a wide array of sensory receptors have now been identified in parasitic worms (Rohde, 1989, 1994a). Some sense organs such as eyes and ears may have been lost by parasitic lineages only to be replaced by more appropriate structures. Brooks and McLennan (1993a) examined the rates of character loss and character innovation in parasitic flatworms, based on the presence or absence of ancestral and derived characters in many taxa. They found that the majority of evolutionary changes in morphological characters were innovations rather than losses. This does not support the view that parasites are structurally simplified and degenerate. This approach would be more convincing if comparisons were made between the rates of character changes in parasitic flatworms and their free-living sister group. Nevertheless, parasites are no more simple than free-living animals, only specialized for a different existence.

The high degree of morphological specialization displayed by parasites also suggests that a transition to parasitism is irreversible. The evolution of parasitism can provide support for Dollo's Law, which states that progress and specialization are unidirectional. Early parasites can evolve a dependence upon a host organism, but once they are dependent on the host there is no going back. In other words, early specialization for a parasitic life commits a lineage forever. The Diplomonadida, a group of flagellated protists containing many species of obligate gut parasites as well as free-living species, suggests a different scenario (Siddall *et al.*, 1993). When the modes of life of the different species are mapped onto their phylogeny, the most parsimonious evolutionary explanation for the observed pattern requires two independent reversals to a free-living lifestyle from parasitic ancestors (Figure 2.1). Are such reversals also possible in metazoan parasites? The nematodes would be the most likely candidates, because they include taxa alternating between free-living and parasitic cycles, but a fully resolved phylogeny will be required to assess the possibility that reversals have also occurred in this group.

2.3 COMPLEX LIFE CYCLES: ACCIDENT OR ADAPTATION?

Organisms that have just completed a transition from a free-living lifestyle to a parasitic existence exploit a single host species. They begin by fine-tuning their exploitation of that host and their methods of transmission to other hosts of that species. During the course of evolution, however, these simple, direct life cycles can become much more complex

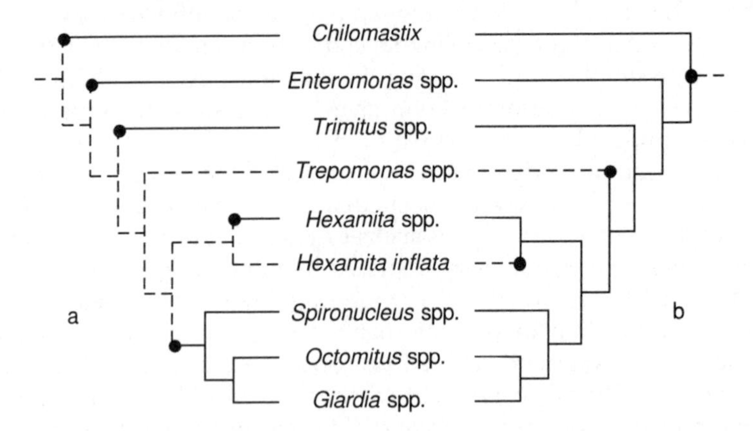

Figure 2.1 Two scenarios for the evolution of parasitism in the protist taxon Diplomonadida. Free-living lineages are indicated by broken lines, parasitic lineages by solid lines; black circles illustrate the proposed transitions between lifestyles. The phylogeny is based on the analysis of 23 ultrastructural characters; *Chilomastix* is the outgroup. In scenario (a) the ancestor was free-living and parasitism arose on five separate occasions. In scenario (b) parasitism was the ancestral state in the Diplomonadida, and there were two reversals to a free-living lifestyle. (Modified from Siddall *et al*., 1993.)

– not because of some intrinsic tendency to evolve toward complexity; parasite evolution is neither more nor less progressive than that of other organisms. Rather, because of historical accidents or because of selective pressures to ensure high transmission efficiency, other host species may be added to the initial cycle and sometimes later dropped. The end result of this process is the panoply of life cycles observed among living parasites, from the most basic and simple ones to fantastically complicated cycles involving many host species, habitats and parasite life stages.

The simplest explanation for the evolution of complex life cycles from simple ones is that new hosts were added to the cycle following historical events that affected the transmission of the parasite or the survival of its host. Selection favoured those parasites capable of adjusting to the new conditions, and the lineages that survived were the ones able to make the best of a difficult new situation. The case of the digeneans offers an example. According to many reconstructions of their evolutionary history (e.g. Shoop, 1988; Gibson and Bray, 1994; Rohde, 1994a), digeneans began as parasites of molluscs. Long after the association between early digeneans and molluscs became established, vertebrates appeared on the scene and began feeding on molluscs. As vertebrates became more abundant, the probability of any mollusc parasites ending up in a vertebrate gut increased. Natural selection would have strongly favoured any parasite with the ability to survive in the gastrointestinal tract of vertebrates, and those able to do so included the ancestors of modern digeneans. As landing in a vertebrate's gut became more and more likely, the parasites made adjustment to their developmental schedule so that the passage to adulthood became linked with the ingestion by a vertebrate host. Other phylogenetic hypotheses suggest that the ancestors of digeneans were parasites of arthropods before they acquired vertebrate hosts, with a switch to molluscan intermediate hosts occurring after the addition of a vertebrate host (see Brooks and McLennan, 1993a). Whatever the exact sequence of events, many features of parasite life cycles can be explained by such adjustments to historical contingencies. Other scenarios trace a slightly different picture, in which the transmission patterns may have become established because they initially benefited the host (Smith Trail, 1980). Such explanations are less parsimonious than those involving only historical accidents, but they also assume that the evolution of parasite life cycles was entirely at the mercy of external factors.

However, viewing such complicated patterns of transmission as mere adjustments to chance events may be simplistic. Bonner (1993) has rightly pointed out that an organism is not just the adult form but the whole life cycle from the fertilized egg to late in the adult life. We may think of a fluke as a flatworm living inside a vertebrate, but natural selection has acted on all stages of its development to shape the life cycle. In fact, it may be more adequate to view the life cycle itself as the unit of selection rather than the parasite. Or, in gene-centred language, the units of selection are the genes coding for the life cycle and not simply for the adult parasite. In this light, we may ask whether life cycles are adaptations rather than mere accidents.

Complex life cycles have evolved independently in several large and distantly related groups of parasitic nematodes (Anderson, 1984; Adamson, 1986; Clarck, 1994; Durette-Desset *et al.*, 1994). Although other scenarios are not impossible, it seems likely that nematodes parasitic in vertebrates began with one-host cycles involving only the vertebrate host. Intermediate hosts, either invertebrates or vertebrates, were added later. In some cases the term paratenic host may be more appropriate; paratenic hosts serve to transfer infective stages from one host to another but parasites do not develop in paratenic hosts. Whether the dependence of the parasite on a particular host is strictly ecological and not also physiological is not important in the present discussion. Some of the advantages of using intermediate or paratenic hosts include protection from the external environment, and the channelling of the parasite toward the final host if the intermediate host is a prey of the definitive host (Anderson, 1984). In the light of these benefits, the common addition of intermediate hosts to the life cycle of nematodes suggests that it was adaptive rather than accidental.

Several other authors have used adaptive arguments to explain the evolution of complex life cycles (Combes, 1995; Ewald, 1995; Morand,

1996a). If fitness benefits can be gained by adding steps to the life cycle, natural selection should favour the parasite individuals able to follow the most rewarding cycle. Ewald (1995) proposed that benefits associated with specialization on different hosts for different resources could have driven the increase in complexity of parasite life cycles. If the fitness of the parasite is greater when one host is used as a food base and another host as an agent of dispersion in time and space, than when a single host is used for both purposes, then a two-host cycle will be favoured. Ewald (1995) supports his idea with data on the virulence of several helminth parasites in their intermediate and definitive hosts, which show that parasites have more severe effects on the intermediate (food base) host than on the definitive (transport) host. The arguments in support of this specialization are not entirely convincing; for instance, few parasites convert tissues from their intermediate hosts into parasite tissues, so that the role of intermediate hosts as food bases is not clear. In any event, data on parasite transmission success and how it varies as a function of the complexity of the life cycle are needed to evaluate life cycles as adaptations.

The only way to assess whether more complex life cycles lead to improved fitness would be to compare the fitness of pairs of related species that differ only in the complexity of their life cycle. An example is provided by two sympatric species of the marine cestode genus *Bothriocephalus*. The two species differ only with respect to their life cycle (Robert *et al.*, 1988). One species, *B. barbatus*, has a two-host life cycle: larval stages live in a planktonic copepod, which is ingested by the final host, a flatfish. The second species, *B. gregarius*, can be transmitted in the same way but usually goes through a paratenic host (a gobiid fish), which feeds on copepods and is itself eaten by the flatfish definitive host. In the Mediterranean Sea where the two cestodes occur, *B. gregarius* is more prevalent and much more abundant in the definitive host than its relative with the simpler life cycle (Figure 2.2). This may suggest that the addition of the paratenic host resulted in greater transmission success. Prevalence and abundance, however, are population measures and cannot be used to assess the success of the individuals on which selection acts. A mathematical model devised for the life cycles of the two species of *Bothriocephalus* suggests that individuals with the more complex life cycles do in fact achieve greater transmission success (Morand *et al.*, 1995). This is due to a larger proportion of infective stages reaching the definitive host when a paratenic host is used. In general, the conditions obtained from mathematical models for the addition of hosts to a life cycle indicate that the additional mortality incurred because of the new transmission step must be compensated by increases in net transmission efficiency or reductions in the mortality of free-living stages (Dobson and Merenlender, 1991).

Many more contrasts between related taxa that differ only in the complexity of their life cycles will be necessary before we know whether or

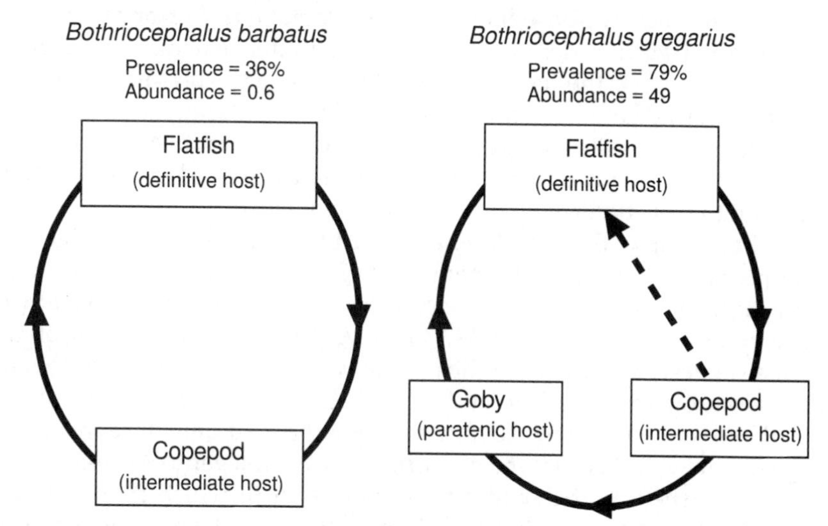

Figure 2.2 Life cycles of the cestodes *Bothriocephalus barbatus* and *B. gregarius*. Values for prevalence and abundance of infection in the definitive host are from populations in the Mediterranean Sea. (Modified from Robert *et al.*, 1988; Morand *et al.*, 1995.)

not complexity is the outcome of selection. General trends associated with evolutionary additions of hosts to life cycles may be difficult to extract from these comparisons, however, because only few of them are possible. Despite the many thousands of living species of helminths displaying a multi-host life cycle, the radiations from which these parasites are issued followed just a handful of transitions from simple to complex life cycles. Acanthocephalans, for instance, are a monophyletic group; all living species and their typical two-host cycles are probably descended from a single common ancestor in which the cycle evolved. Complex cycles have had a few independent origins among protozoans, nematodes and platyhelminths parasitic in animals (Anderson, 1984; Adamson, 1986; Barta, 1989; Brooks and McLennan, 1993a; Rohde, 1996). The phylogenetic rarity of these changes in the complexity of life cycles may prevent any robust trend from emerging. Comparisons between individuals of the same species that differ in their life cycles would be even better though these are fewer, as only a few species have facultative steps in their life cycle.

If the addition of hosts to simple life cycles can be advantageous in some cases, surely the loss of hosts from complex cycles could also be advantageous in other situations. For example, blood flukes of the families Spirorchidae and Schistosomatidae display a two-host cycle most probably derived from an ancestral three-host cycle, typical of the majority of digenean groups (Shoop, 1988; Brooks and McLennan, 1993a). Is the abbreviated two-host life cycle an adaptation? Cercariae of blood flukes penetrate the definitive host directly after leaving their mollusc intermediate host; the 'current' definitive host probably used to be the second intermediate host in the ancestral life cycle, and became the host in which the parasite matures after the original definitive host was lost (Shoop, 1988). If this return to a simpler life cycle is adaptive we might expect it to have occurred in other digenean lineages.

A shortening of the life cycle has indeed taken place in other digenean groups. Some species of the genus *Alloglossidium* have a normal three-host digenean life cycle, but other species in the same genus have a two-host cycle (Carney and Brooks, 1991). A phylogenetic analysis of the genus suggests that the three-host cycle was the ancestral condition and that species with a two-host cycle are all derived from a common ancestor, the loss of the vertebrate host having happened only once in the lineage (Figure 2.3). Also, in one or more species of the genus *Mesostephanus*, miracidia penetrate the snail intermediate host in which they multiply asexually. Instead of producing cercariae (the next stage in the life cycle responsible for infecting the fish serving as second intermediate host), they can produce and release new miracidia, which can reinfect other snails (Barker and Cribb, 1993). The parasite skips two hosts and one round of sexual reproduction each time it does this. The circumstances triggering this unusual development are unknown, but this example and the previous one show that shortened life cycles can evolve.

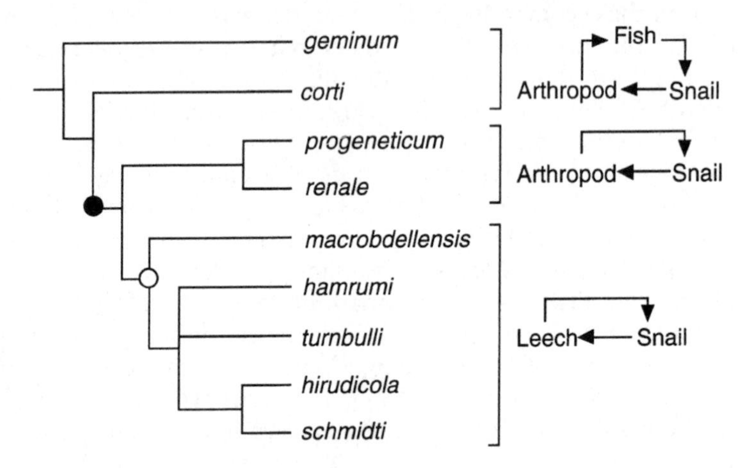

Figure 2.3 Phylogeny of species in the digenean genus *Alloglossidium* showing the evolutionary changes in life cycle patterns. The three-host cycle is the ancestral condition and is still present in two species. The fish host was lost only once in the evolution of the lineage (black circle), and there was one host switch in which the arthropod host was replaced by a leech (open circle). Worms using only two hosts reach adulthood in either the arthropod or the leech. (Modified from Carney and Brooks, 1991.)

If a change in the complexity of the life cycle is adaptive, it may lead to higher rates of speciation because it could allow the colonization of new niches and lower rates of extinction. Brooks and McLennan (1993b) compared the species richness of the major groups of parasitic platyhelminths. They concluded that of the three groups (Digenea, Monogenea and Eucestoda) displaying an independently derived high level of species richness, an adaptive radiation has only occurred in the Monogenea. They attributed this to the appearance of a key innovation involving the loss of one host from the life cycle and the reversal to a direct one-host cycle. This would support the claim that changes to the complexity of the life cycle are adaptive and are driven by selection. However, the conclusions of Brooks and McLennan (1993b) are valid only if they used the correct phylogeny to derive contrasts in richness among platyhelminth taxa, and if their estimates of species richness are accurate. Rohde (1996) has questioned the analysis of Brooks and McLennan (1993b), and has convincingly shown that the results are entirely dependent on a phylogenetic hypothesis that may be flawed. As for estimates of species diversity, there are indications that we have identified only a fraction of existing species (Poulin, 1996a; Rohde, 1996), so that any attempt to compare the richness of lineages with different life cycles may be premature.

In conclusion, the adaptiveness of parasite life cycles will prove difficult to demonstrate. The possibility that they are adaptive is enticing, and at least forces us to consider alternatives to classical ideas founded solely on the constraints of history. A final example illustrates how challenging old ideas brings a fresh new perspective on parasite evolution. The larvae of many nematodes that are parasitic in vertebrates undergo extensive migrations through the tissues of their host. Some migrating species infect the host by penetrating the skin, and have no choice but to migrate through the host to reach the host's gut, where they spend their adult life. Other migrating species, however, are acquired by ingestion; they start their migration from the host's gut only to end up back there. The prevailing explanation for migration in these latter species is that it is an evolutionary legacy, inherited from a skin-penetrating ancestor (see for example Adamson, 1986). Comparisons between migrating nematode taxa and their non-migrating sister group show that migrating parasites are significantly larger, and presumably more fecund, than their non-migrating relatives (Read and Skorping, 1995). Migrating worms achieve faster growth rates, possibly because of some nutritional advantage over worms developing entirely in the gut. In the strongylid nematodes, skin penetration and tissue migration appears to be the ancestral condition (Sukhdeo *et al.*, 1997); the loss of skin penetration and the retention of migration as the nematodes adopted an oral infection route suggest that migration is a legacy, but it could have been maintained and favoured by selection if it accelerated growth. Thus, nematode migrations appear to be more than just vestigial behaviour patterns, and perhaps a similar adaptive explanation applies to parasite life cycles in general.

2.4 EVOLUTIONARY CONSEQUENCES OF COMPLEX LIFE CYCLES

Whether or not complex life cycles are the products of selection, they have in turn imposed selective pressures on parasites and lead to drastic adjustments in the parasites' biology. In the course of a complex life cycle, parasites will inhabit taxonomically unrelated hosts, visit different physical habitats, and need means of moving from one host or habitat to the next. They will assume widely different shapes, each fitted to a given part of the cycle. An adult digenean, a miracidium and a cercaria are very distinct in morphology, but are all vehicles of the same genetic information. At each stage of the cycle, selection has favoured morphologically different carriers of the parasite's genes in response to pressures from the cycle itself.

The addition of a new step in transmission may or may not lead to fitness benefits, but it entails a new set of challenges that must be met by the parasite. Each stage presents an array of adaptations aimed at facilitating the completion of that stage (Figure 2.4). Some of the same adaptations are also expected in parasites with simple one-host cycles, and these parasites will also be used as examples here. From the moment eggs are released by the adult parasite in the definitive host, a cascade of events must take place in the proper sequence for the successful return of the offspring to the right definitive host. The constraints and pressures on egg production by adult parasites will be dealt with in detail in Chapter 4. This section focuses on the hurdles and challenges facing parasites from the moment eggs are released by the adult, and on how these have been overcome. Adaptations intrinsic to parasitism itself, such as immune evasion or the extraction of nutrients from hosts, are not considered in detail here; the focus is on those evolved in response to the increased complexity of life cycles. Combes (1995) provides a more comprehensive review of these and other adaptations to complex life cycles.

At each stage in the cycle, selection favours those parasites better at infecting the next host in the cycle, the target host. Because target hosts, whatever they are, are patchily distributed in space and time, the likelihood of finding them can be increased with a precisely timed release of eggs or infective stages. The cestode *Triaenophorus crassus*, for example, stores up to a few million eggs in its body during its year-long life in the pike, the cestode's definitive host. Pike enter shallow water only once a year, in order to spawn. The cestode's first intermediate host is a planktonic copepod inhabiting shallow water. Egg release by the cestode coincides with the pike's spawning activity: after a year during which eggs are accumulated, they are released in a brief instant the moment the parasite finds itself in the microhabitat of its next target host (Shostak and Dick, 1989). An even more striking example of well-timed egg release comes from some polystomatid monogeneans, in which the temporal window of opportunity for transmission is even narrower. Polystomatids, like other monogeneans, are transmitted by a swimming ciliated larva; however, unlike other monogeneans, many polystomatids exploit vertebrates that do not spend their entire life in water. One species, *Pseudodiplorchis americanus*, lives in the urinary bladder of spadefoot toads in Arizona (Tinsley, 1990). The toads live in deserts and hibernate in burrows for 9–10 months per year. They become active following the annual summer rains, and spawn in temporary pools. Spawning takes place over one to three nights, and individual toads may spend less than 24 hours per year in water. The parasite has synchronized its egg release with these brief visits to water, abruptly expelling a whole year's production of offspring into the host's urine when the host enters water.

The timing of egg release with the period of greatest availability of the target host is only the first step toward ensuring that a target host will be reached. In many parasites eggs do not hatch until they have been ingested by the target host. As the time between release from the definitive host and ingestion by the target host may be long, selection has often favoured eggs that can remain viable for very long periods. Nematode eggs, for instance, possess a thick, resistant shell that protects the larvae against environmental hazards and allows them to survive for several years in some cases (Wharton, 1986).

In other parasites, eggs hatch to release infective larvae that must locate and infect the target host. Because parasite larvae typically do not feed and are therefore short-lived, hatching of the eggs must also be synchronized with the presence of target hosts if these are transients and if their availability is variable. Again, monogeneans provide good examples. Hatching of monogenean eggs is typically finely tuned either to environmental cycles or to stimuli of direct host origins (Kearn, 1986; Tinsley, 1990). In species exploiting hosts with a clear-cut circadian behaviour pattern which leads to day-night differences in their susceptibility to infection, egg hatching often shows circadian rhythms timed with photoperiod. In other species exploiting hosts with unpredictable availability, eggs remain viable for long periods and hatching is triggered only by specific cues. These include chemical substances specific to the host mucus, as well as less specific stimuli such as passing shadows or physical disturbances in the surrounding water. Such precise hatching synchrony illustrates how selection has finely tuned the match between the parasite's biology and that of the host, as a response from pressures generated by the life cycle itself.

The release of eggs by the adult parasite in the definitive host may not be the only place in the cycle where infective stages are released to find a target host. In digeneans, the life cycle also includes the release of cercariae from the first (mollusc) intermediate host and the subsequent infection of a second intermediate host (Figure 2.4). Here the life cycle also favours the parasite genotypes which correspond to a perfect match between the timing of cercarial release and the susceptibility of the target host. Cercariae are typically short-lived, having a life span of a few to several hours (Combes *et al.*, 1994). The main factor synchronizing cercarial shedding from mollusc hosts is the photoperiod. Patterns of emergence, however, vary across species and appear to correspond to the patterns of activity of the target host (e.g. Théron, 1984; Combes *et al.*, 1994). Among schistosomes, cercarial emergence coincides with the periods during which target hosts are most likely to visit water (Figure 2.5). Other environmental influences, such as predator avoidance, may sometimes play a role in the evolution of the diurnal pattern of cercarial emergence (Shostak and Esch, 1990). But in the majority of cases, the precise fit between the activity of the target host and the timing of cercarial emergence is strong evidence that emergence patterns are adaptations for transmission. In addition, experimental crosses between populations of the same schistosome species that differ in the timing of cercarial emergence indicate that cercarial release is under genetic control (Combes *et al.*, 1994).

Maximizing temporal overlap with the target host is one of two types of adaptation required by the infective stages of a parasite in order to complete their phase of the life cycle. The other challenge is to find the host in space (Figure 2.4). The mobile infective stages of digeneans display a range of responses to simple cues that facilitate host location (Saladin, 1979; Sukhdeo and Mettrick, 1987; Combes *et al.*, 1994). Infective stages can either respond to cues from the physical environment that bring them within the area inhabited by the target host, or on a smaller scale they can respond to specific cues produced by the target host (Combes, 1991a; Combes *et al.*, 1994). Miracidia searching for snail hosts are relatively specific in their choice of host species. They can identify specific macromolecules released by their hosts and respond by altering their swimming behaviour: they increase the frequency of turning which maintains them in the small area where host chemicals have been detected (Haas *et al.*, 1995). Cercarial host location seems to operate mostly on a larger scale. Cercariae of different species, for instance, show different responses to light or gravity and consequently migrate up or down the water column in an attempt to match the spatial distribution of their target host (Figure 2.6). On a finer spatial scale, not all cercariae appear to be able to respond to host chemicals (Combes *et al.*, 1994; Haas, 1994; Haas *et al.*, 1995). Cercariae often have a broader range of acceptable host species than miracidia, but the lack of specific host-recognition mechanisms in many species results in mistakes and the death of the cer-

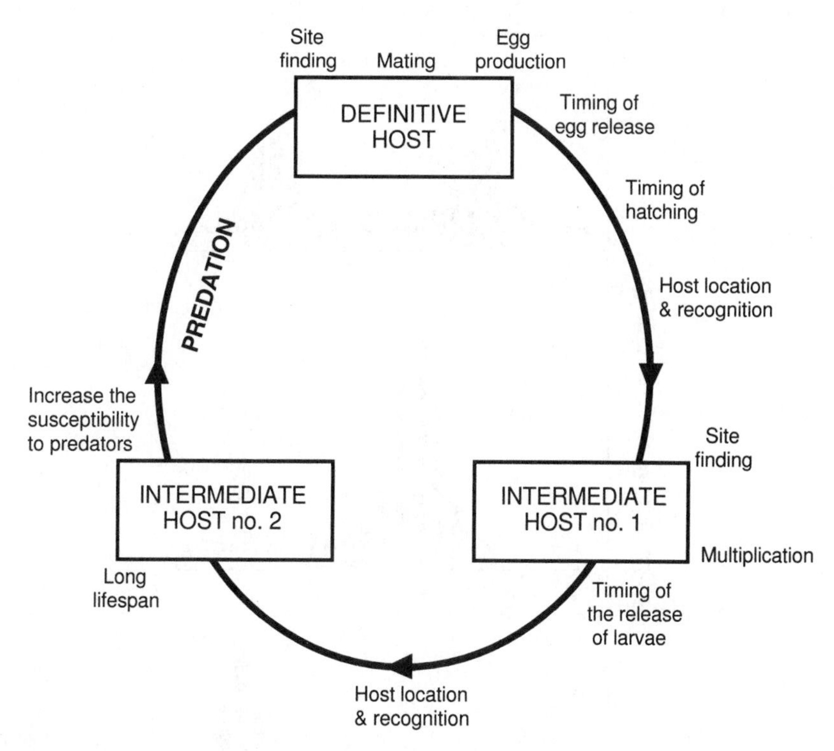

Figure 2.4 Generalized three-host life cycle, based on that of a digenean. Adaptations favoured by the cycle itself are indicated at each step in the transmission from one host to the other.

cariae. The dermatitis caused by avian schistosomes accidentally penetrating human skin is an example of such mistakes. That the ability to recognize suitable hosts has not evolved in cercariae of all species is therefore surprising.

At each transmission stage in the life cycle, the probability of finding a suitable host is very small indeed. The addition of new steps in the cycle has apparently stacked the odds against the completion of the cycle by any individual larva. Whether these new steps were added by accidents or through selection, the pressure has been on parasites to come up with ways to compensate for the heavy losses incurred during successive transmission events. In a cycle such as that of Figure 2.4, with two stages involving the release of propagules and the infection of hosts by mobile larvae, larval mortality can make the completion of the cycle unlikely. If the first intermediate host is used only as a food source for the transformation of the first mobile larva into a single new larva for the next stage, and if the probability of success in both transmission events is 0.001, then

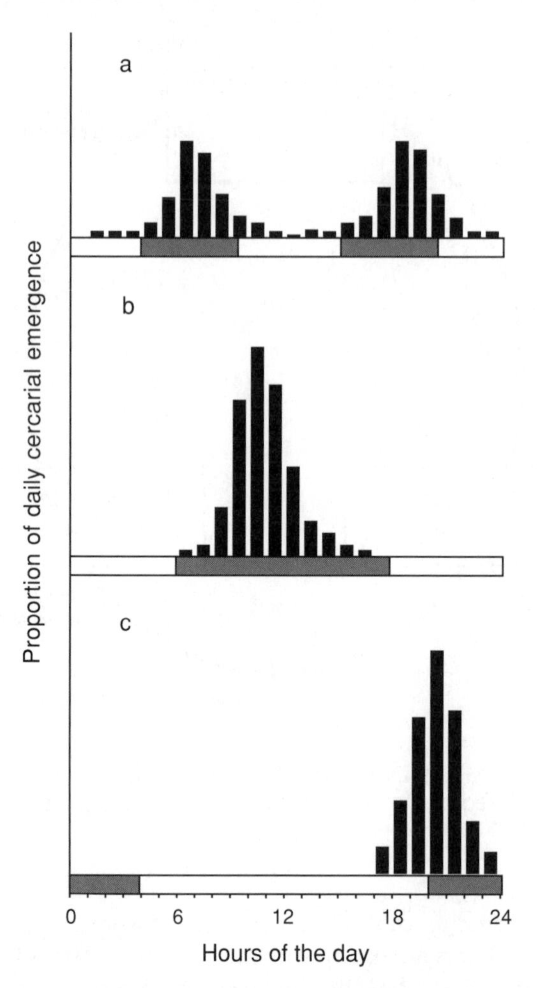

Figure 2.5 Patterns of cercarial emergence in three species of schistosomes with different target hosts. Black vertical columns represent the relative number of cercariae released for each hour of the day; approximate periods of host visits to water are indicated by shading in the horizontal bars. The species are: (a) *Schistosoma margrebowiei*, infecting African ungulates; (b) *S. mansoni*, infecting humans; and (c) *S. rodhaini*, infecting rats. (Modified from Combes *et al.*, 1994.)

out of a million eggs produced by the adult, on average only a single one would reach the second definitive host. And the cycle would not yet be completed, as the parasite must survive, grow and mature successfully inside the host. Although these numbers are arbitrary, they illustrate the sort of odds faced by parasites with complex life cycles.

There are at least three obvious solutions to this problem, and parasites typically resort to more than one of them. Firstly, selection may

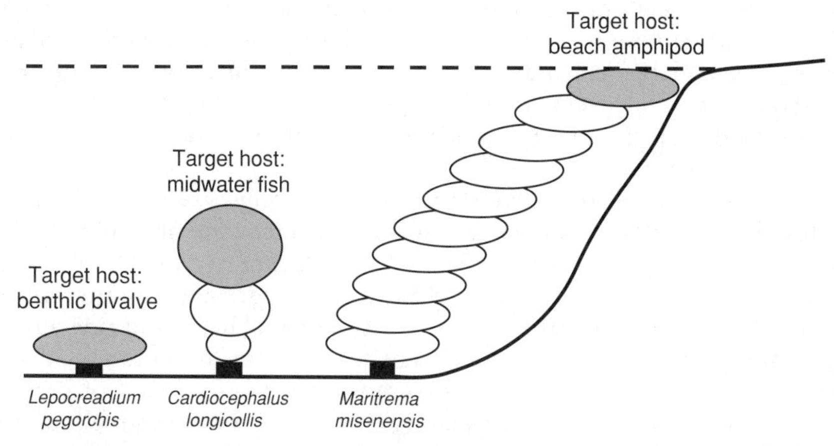

Figure 2.6 Dispersion patterns of cercariae of three species of digeneans after their emergence from their benthic molluscan host in a marine littoral habitat. (Modified from Combes *et al.*, 1994.)

favour the adult which produces more eggs; this is seen as a typical adaptation of parasitic helminths and will be discussed in Chapter 4. Secondly, selection could increase the success rate and favour the production of larvae better equipped for survival and host location. The costs of producing such larvae may impose limits on how many could be produced; again, Chapter 4 will return to such potential trade-offs in life history strategies. Thirdly, selection could favour the production of a fresh new wave of infective stages by the original larvae that were successful in the first infection event. This strategy is a characteristic of digeneans: each miracidium successful at infecting a mollusc intermediate host will multiply asexually to produce large numbers of cercariae ready to embark on the next leg of the cycle. For example, in schistosomes, thousands and even tens or hundreds of thousands of cercariae may be produced from one miracidium (Loker, 1983). Similar numbers are observed in other digenean groups (e.g. Shostak and Esch, 1990; Combes, 1995). Despite the obvious benefit of asexual multiplication, it is difficult to evaluate whether it is a true adaptation and arose through selection, or whether it originated in some other way and is only coincidentally advantageous. As all digeneans share this trait, it is most probably the outcome of a single evolutionary event, i.e. it appeared once in the ancestor of all digeneans. The independent appearance of a trait in several lineages under similar selective pressures would provide stronger evidence of adaptive convergence. Digeneans are not the only taxa capable of asexual multiplication in intermediate hosts; the trait has evolved independently in several cestode lineages (Moore, 1981; Moore and Brooks, 1987). It is therefore likely that this amplification of numbers is a trait strongly favoured by selection but restricted to lineages in which it can evolve because of sufficient genetic variation. The nature of the intermediate host used can also constrain the evolution of asexual amplification; for instance, many arthropods used as intermediate hosts by helminths may not provide sufficient resources for asexual multiplication, or may not be able to survive it.

The use of mobile infective stages is not the only way in which parasites are transmitted from one host to another. Predation of an intermediate host by the definitive host (Figure 2.4) is common in the life cycles of parasitic protozoans, digeneans, cestodes, nematodes and acanthocephalans. Many of these parasites exploit existing food chains and ride along predator-prey interactions, a fact that leads many to believe that life cycles are merely the accidental by-products of the evolution of predation. In any event, in cycles depending on such rare events as the capture of infected prey by suitable predators, selection could favour at least two different strategies. It could favour either parasites capable of increasing the frequency of such events, or those patient enough to wait for the rare events to happen (Poulin, 1994a). Many parasites induce modifications to the appearance or behaviour of their intermediate hosts. These modifications may often be adaptive manipulations by the parasite of the intermediate host's susceptibility to predation by the definitive host (see Chapter 5). Alternately, parasites could accept the rarity of predation and wait patiently for it to happen by increasing their life span in the intermediate host. For example, plerocercoids of the cestode *Triaenophorus crassus* can survive for several years in their fish intermediate host waiting to be ingested by a predatory pike, the definitive host (Rosen and Dick, 1984). Other cestodes not known to alter the behaviour of their intermediate hosts can survive in them for years (Moore, 1981). In these parasites the life cycle is interrupted by long periods, without growth or development, spent waiting for an improbable event. Patience and manipulation are possibly alternative strategies, either incompatible or too costly to be both selected in the same parasite (Poulin, 1994a; Chapter 5). This again illustrates the sort of dilemma imposed on parasites by complex life cycles.

Reaching the definitive host does not signal the end of the cycle; the parasite still has two tasks to achieve. First, it must find its site of infection, the one for which its adult phenotype is designed. This site is often very specific, and may have very precise boundaries even within a large organ such as the vertebrate intestine. Fitness of the parasites is dependent on how close they come to the optimal attachment site: for example, the precise location of worms in the host's intestine determines the likelihood of mating and subsequent fecundity in the acanthocephalan *Moniliformis moniliformis* (Lawlor *et al.*, 1990) and female fecundity in the nematode *Trichinella spiralis* (Sukhdeo, 1990a). During the evolution of the life cycle, the route of entry into the definitive host may have been changed entirely. For instance, parasites that used to penetrate the host through the skin may have evolved to use the oral route of infection. Changes like this during the evolution of the life cycle would have put great pressures on parasites to adjust their way of reaching their final site of infection. Sukhdeo (1990b) suggested that the host internal environment is much more predictable than the environment at large, because all conspecific hosts are almost identical in construction and function, the same organ secreting the same chemicals in all hosts, etc. In these predictable conditions, any behaviour increasing the chance of arriving at the correct site of infection would be favoured and would quickly spread to fixation through the parasite population (Sukhdeo, 1990b; Sukhdeo and Sukhdeo, 1994). The rapid evolution of fixed, simple behaviour patterns could make adjustments in site finding relatively easy, following alterations to the life cycle affecting the route of entry into the definitive host.

The final tasks necessary to complete the cycle are finding a mate and the act of mating. Because reaching the definitive host after going through the whole cycle is a relatively rare event, the co-occurrence and physical encounter in the same host at the same time of a male and a female parasite of the same species is unusual in most situations. Several possible solutions exist. Hermaphrodism is an ancestral trait of platyhelminths; these parasites do not need to find a mate of the opposite sex, any conspecific will do. Self-fertilization is also common, especially in tapeworms, and eliminates the need to find a partner altogether. In nematodes and ticks, some ancestrally dioecious lineages have evolved toward parthenogenesis and eliminated the need for eggs to be fertilized. However, the majority of ectoparasitic arthropods, endoparasitic nematodes and acanthocephalans have separate sexes that must meet for egg fertilization. Schistosome digeneans are also dioecious, but unlike nematodes or acanthocephalans, dioecy is a derived trait and evolved from a hermaphroditic condition (Basch, 1990). Why did selection favour a return to dioecy in schistosomes? It may have something to do with a division of labour between the sexes leading to a higher fitness for both participants than simple hermaphrodism. Note that the typical (but not universal) mating pattern of schistosomes is one of life-long monogamy, with females and males remaining physically attached for life. Recently, it has been shown that parthenogenesis has evolved, possibly more than once, in schistosomes (Jourdane *et al.*, 1995). When a female schistosome is paired with a male from a distantly related species, offspring are produced through parthenogenesis; in at least one species, *Schistosoma douthitti*, parthenogenesis occurs regularly even when conspecific males are available, possibly an advantage when population density is low. It is intriguing that schistosomes, after switching to dioecy, show a tendency to revert to a mode of reproduction not involving pairing. Although this latest ability appears advantageous, we can only speculate that the life-cycle conditions under which dioecy was favoured were different from the conditions under which parthenogenesis has appeared.

Everything in the biology of parasites, from larval development to adult reproduction, has been shaped by the nature of the life cycle. The complexity of the transmission process from a definitive host and back to another definitive host has forced parasites to adopt many guises, each a different manifestation of the same genotype. This section has presented only a brief survey of the many influences of the life cycle, but is intended to demonstrate the key role of the cycle in parasite evolution.

2.5 SUMMARY

From simple beginnings, parasite lineages have evolved complex adaptations to their way of life. The ephemeral nature of the host as a habitat has forced parasites to find ways of constantly colonizing new hosts. Individuals that are better at doing so because they go through additional steps in their life cycle are favoured by selection, and thus evolve complex life cycles. In turn, these products of selection act as strong selective pressures leading to adaptive adjustments in all aspects of the parasite's biology.

The variety of parasite life cycles observed today is evidence that parasite evolution does not follow a single road. There are, however, striking similarities between the life cycles of very distantly related groups in which parasitism had independent origins. In endoparasitic platyhelminths, nematodes and acanthocephalans, complex life cycles involving transmission by predation from an invertebrate intermediate host to a vertebrate definitive host are extremely common. Adaptations to such a life cycle also appear to have evolved in parallel in these three groups, providing good examples of convergent adaptations. Therefore ancient events in the phylogeny of parasites can commit a lineage to a particular evolutionary path, and make difficult the distinction between constraint and adaptation. Contrasts between the transmission rates of sister taxa differing only in one aspect of the life cycle offer the only way to achieve this distinction.

Host specificity 3

Over evolutionary time, parasites have added hosts to their life cycles by adding steps to the cycle and thus increasing its complexity. Hosts can also be added to the life cycle in parallel rather than in series (Combes, 1991b, 1995). The spectrum of potential hosts that can be used at any step in the cycle can be broadened without an increase in the number of steps. Instead, selection simply adds alternative pathways through the cycle. A complex life cycle allows parasites to specialize on two or more hosts by partitioning specialization to different times during the cycle (Thompson, 1994); however, at each step in the temporal sequence of the cycle, the specialization can be relaxed to include more than one host species. This chapter will extend the previous chapter's discussion of the evolution of complexity in parasite life cycles by discussing the evolution of host specificity at different stages in the life cycle, and the wide differences in specificity displayed by extant parasites.

3.1 MEASURING HOST SPECIFICITY

The term host specificity has been used in different contexts and has many interpretations. The most common meaning is adopted here, and host specificity is defined as the extent to which a parasite taxon is restricted in the number of host species used at a given stage in the life cycle. Highly host-specific parasites are restricted to one host species and specificity declines as the number of suitable host species increases.

Because specificity is measured as the number of host species of a parasite, it can be estimated by summing up the number of known host species from published records of parasite occurrence. There is an obvious danger in using these estimates, however. Consider two parasite species – the first which has been the subject of a single survey, in which it was described from one host species at one location; and the second which has been regularly reported from several populations of the same host species over a large geographical area. Both these parasite species have a single known host species. But can we be sure that the first species is not in fact exploiting a broad range of host species in which it is yet to be found? High host specificity can be an artefact of inadequate sampling. Among species of parasites of freshwater fish, sampling effort explains much of the variability in host specificity (Poulin, 1992a, 1997a): the number of known host species is strongly, positively correlated with the number of times a parasite species has been recorded (Figure 3.1). The same is true among tick species, and the distinction between highly specific and less specific ticks may really be a distinction between rarely and frequently collected species (Klompen *et al.*, 1996). If the number of host species used is characteristically underestimated in poorly studied parasite species, then it is not a very adequate measure of host specificity. Corrections for sampling effort, however, can make this measure more reliable (Poulin, 1992a).

Another problem in using lists of published records is that this method does not provide an accurate measure of the specificity of parasites in one population, only that of their species as a whole. If a species of parasite is known to exploit seven host species, there is no reason to believe that individuals in a given population of that species are capable of infecting all seven hosts. Members of the population may be adapted only to locally available host species, and thus be more host specific than their species as a whole.

Incorrect species identification can also influence estimates of host specificity. On the one hand, a species of parasite known to exploit *n* host species in a given area can in fact prove to be a complex of *n* species of superficially identical, highly host-specific parasites, once accurate molecular analyses are performed (see examples in Combes, 1995; Thompson and Lymbery, 1996). On the other hand, what appear to be *n* related species of parasites exploiting *n* different host species can prove to be a single parasite species with low host specificity in which the phenotype is influenced by the identity of the host species, with a resulting confusion in taxonomy. What follows ignores the above problems by assuming they can be accounted for.

Some authors have referred to the number of host species used as the host range, and proposed that true measures of specificity should take into account how heavily and how frequently the various host species are infected by a given parasite (Lymbery, 1989; Rohde, 1993, 1994b). While the traditional definition of host specificity is adhered to here, there is no doubt that information on whether a parasite species utilizes its various host species equally, or whether it concentrates on only one, would also be valuable. Rohde (1980) developed an index of specificity, *S*, based on the number of parasite individuals found in each host species.

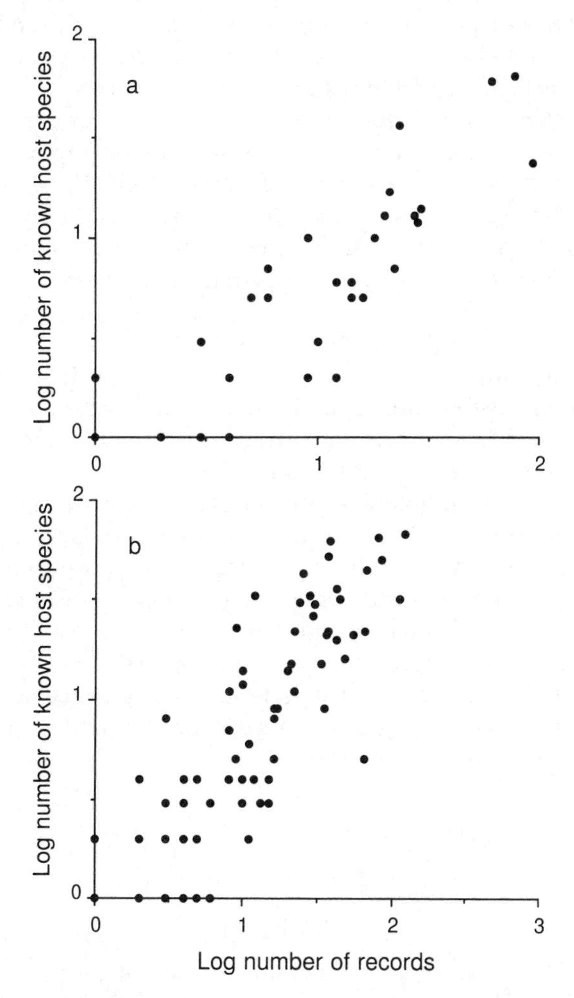

Figure 3.1 Relationship between the number of known host species and the number of published records mentioning a parasite genus among metazoan genera of ectoparasites (a) and endoparasites (b) of Canadian freshwater fish. More hosts tend to be known for intensely-studied parasite species. (Data from Poulin, 1997a.)

The index can be used only for a large sample including many species of hosts collected at the same time and location, and is computed as follows:

$$S = \frac{\sum(x_i / n_i h_i)}{\sum(x_i / n_i)}$$

where x_i is the number of parasite individuals on the ith host species, n_i is the number of host individuals examined of the ith host species, x_i / n_i is the abundance of parasites on host species *i*, and h_i is the rank of host species *i* (the host species with greatest abundance of parasites has rank 1). Rohde (1980, 1993, 1994b) suggested that the index could be modified to use prevalence of infection instead of abundance as a measure of host utilization by parasites. Prevalence is a measure of how widespread the parasite population is in a sample of hosts; it is not a measure of how individual parasites are transmitted to their hosts. As abundance is a better indicator of the success of individual parasites at infecting hosts of different species, it is the most appropriate measure to use in the computation of Rohde's index.

The value of the index tends toward one when the parasites concentrate predominantly in one host species, and is expected to tend toward zero when parasites are equally distributed among host species. However, when host species are used evenly, the value of the index approaches zero only when the number of host species included in the computations is much greater than 20 (Figure 3.2). The majority of parasite species utilize fewer than 10 host species and for such species a high value of *S* does not necessarily imply that parasites concentrate on a single host species. Rohde's index may therefore be unreliable for comparisons of parasite species using different numbers of host species, because similar scores may represent different patterns of host utilization. Nevertheless, it is the only such index currently available. A modified index, in which the value of *S* is corrected for the number of host species used, would be more appropriate.

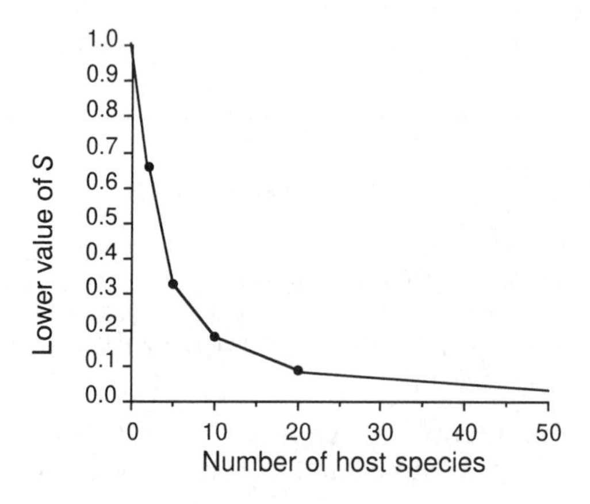

Figure 3.2 Changes in the minimum possible value of Rohde's (1980) index of specificity, *S*, as a function of the number of host species utilized by a parasite. The lower bound for *S* was obtained in simulations in which all host species were assigned identical, average ranks according to how many species there were. The upper value of *S* is always 1, independently of the number of host species. Thus the range of values that *S* can take increases with the number of host species utilized.

Most parasites show bias in the way they utilize suitable host species (Figure 3.3) and some index of host preferences would complement measures of host specificity. Ranking of host species according to parasite abundance is one way to assess which hosts are preferred among the spectrum of suitable species (Figure 3.3). Robust estimates of abundance require the examination of many hosts, and rankings of host species are only meaningful if all species have been sampled equally. Even when host species are well sampled, however, ranking them with respect to parasite abundance in order to determine which species are preferred may not be informative, because different host species are not equally available to parasites. The relative abundance of hosts of different species may determine to what extent they are used by parasites and mask true preferences if any. For instance, if parasites are twice as abundant on host species A than on host species B, but host species A is twice as common in the environment as host species B, we cannot conclude that A is preferred. Proper indices of host preferences should take into account both use and availability of hosts (Lymbery, 1989), and are therefore difficult to use in natural situations. An alternative method of assessing preferences could involve quantifying development rate of the parasite on different species of hosts (Kennedy, 1975).

Measuring host specificity is not as straightforward as it may appear at first glance. Detailed analyses of patterns of host-species use are complicated by the lack of information on the relative abundance of different host species. Even simple lists of known host species may be incomplete because of an unequal sampling effort across parasite species. These

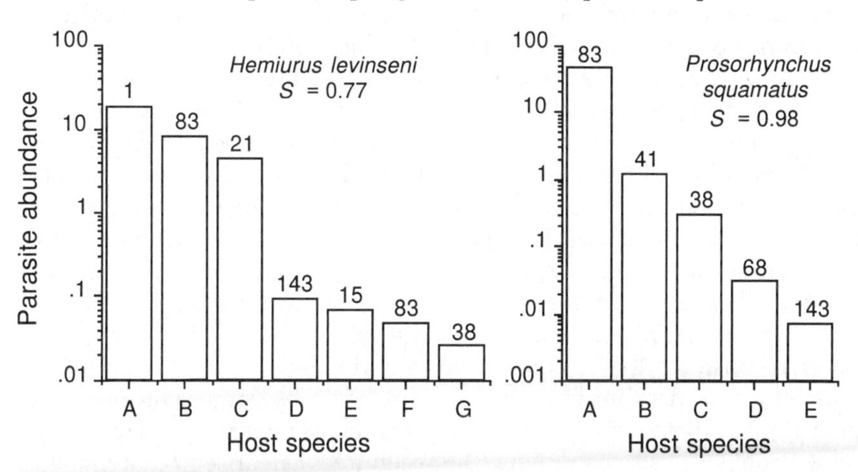

Figure 3.3 Parasite abundance (total number of parasites found divided by number of hosts examined) in different host species for two species of digeneans parasitic in marine fish. Note that abundance is on a log scale. The numbers of fish examined in each host species are indicated above the bars, and so is *S*, Rohde's index of host specificity. (Data from Rohde, 1994b.)

problems should be kept in mind when reading the section on observed patterns of specificity.

3.2 HOST–PARASITE COEVOLUTION AND HOST SPECIFICITY

To understand why some parasites are very host specific and others are not, we must first understand how parasites and hosts have coevolved since the origin of their association. Recent years have seen much progress in our efforts to reconstruct the phylogeny of parasites based on morphological or molecular information. The many available parasite phylogenies have made possible comparisons between the phylogeny of a group of parasites and that of their hosts. Differences and similarities between host and parasite phylogenies can shed light on the history of their association. Developments in the field of host-parasite coevolution as it pertains to the evolution of host specificity are briefly reviewed here.

Consider a speciation event taking place in an ancestral host population harbouring one species of parasite. If the barrier to gene flow that isolates the two allopatric subpopulations of hosts also prevents gene flow between the two newly created subpopulations of parasites, then the parasite will cospeciate with its host. If this process of 'association by descent' is repeated several times in the daughter species, it will result in *n* species of hosts and *n* species of parasites. Each parasite species will be strictly host specific and the phylogeny of the parasite species will be a mirror image of that of their host species (Figure 3.4a). The perfect congruence of host and parasite phylogenies is a pattern known to parasitologists as Fahrenholz's Rule, and it can serve as a null model of host-parasite coevolution against which other evolutionary scenarios can be tested. Typically, host and parasite phylogenies will not be entirely congruent and parasites will not show strict host specificity (Figure 3.4b). The occurrence of one parasite species on more than one host species can result from the continuation of gene flow between the parasite populations on different hosts, and the maintenance of a single parasite species following host speciation. It can also result from host switching, or the colonization of new host species. The opportunities for host switching and the frequency of such events in the phylogeny of parasites will determine the number of host species they can exploit.

To reconstruct the history of cospeciation with the host and host switching in a parasite lineage, we must begin by comparing host and parasite phylogenetic trees and assessing their congruence. This poses some problems. False congruence between trees can arise from a sequential colonization of hosts by parasites which coincidentally mirrors the pattern of host radiation (Brooks and McLennan, 1991). False incongruence between host and parasite trees is also possible, and can result from duplications and losses of parasites independent of host phylogeny

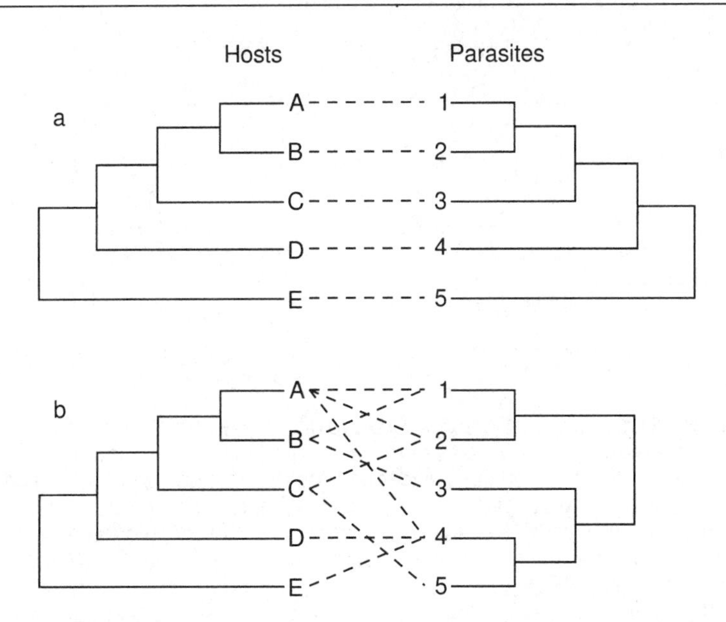

Figure 3.4 Phylogenetic trees of hypothetical host and parasite species. Broken lines indicate host-parasite associations. (a), Host and parasite trees are perfectly congruent, and parasites are strictly host specific. (b), Host and parasite trees are incongruent, and the host specificity of parasites varies among species. (Modified from Barker, 1991.)

(Page, 1993; Paterson and Gray, 1996). The absence of parasite species from host species in which they should occur if both lineages had cospeciated can be the result of parasite extinction following a cospeciation event (Figure 3.5). Parasites may also simply have been absent from one of the two subpopulations of hosts during a host speciation event. The aggregated distribution of parasites among hosts (Chapter 6) can result in a founder host population being parasite-free (Paterson and Gray, 1996). The duplication of a parasite lineage without a corresponding speciation in the host lineage will create minor discrepancies between the branching patterns of host and parasite phylogenies (Figure 3.5). Sampling error can also lead to false incongruence between the trees. If the abundance of parasites is very low, they may escape detection (Paterson and Gray, 1996). In addition, the presence of members of a parasite species on an unusual host species may be a case of 'straggling' (i.e. the accidental and temporary occurrence of a parasite in or on the wrong host), rather than either a successful host-switching event or one in progress (Rózsa, 1993).

There are at present two main analytical methods of measuring the congruence between host and parasite phylogenies. One of them, Brooks' parsimony analysis (BPA), treats parasites and their phylogeny as host characters and then compares the most parsimonious host

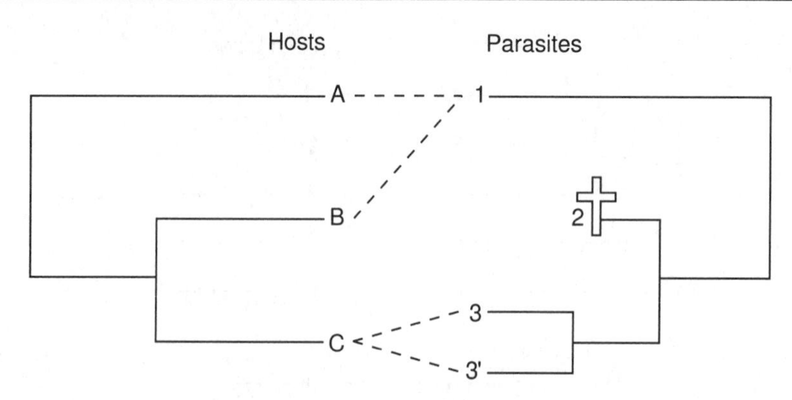

Figure 3.5 Illustration of some possible evolutionary events that can lead to incongruence between host and parasite phylogenies: (i) Host switching, in which parasite lineage 1 evolved on host lineage A before colonizing host lineage B; (ii) extinction, such as what happened to parasite lineage 2 during its evolution on host lineage B; and (iii) intrahost speciation, such as the duplication of parasite lineage 3 without a corresponding speciation event in host lineage C. (Modified from Paterson and Gray, 1996.)

cladogram derived from the parasite data with the original host phylogeny (Brooks, 1988; Brooks and McLennan, 1991). Incongruence is taken as evidence for one or more host-switching events, which can lead to an inflated estimate of the number of host switches (Paterson and Gray, 1996). The other method, reconciliation analysis, contrasts independently derived host and parasite phylogenies, and determines whether their congruence is greater than expected by chance (Page, 1990, 1993). Reconciliation analysis assumes that host switching has not occurred, and explains incongruence with parasite extinctions or duplications such that all incongruence is false and cospeciation is the rule. Despite the different assumptions behind the two approaches and their inherent biases, both usually lead to similar general conclusions when applied to the same data set (Hoberg *et al.*, 1996; Paterson and Gray, 1996). Recent modifications to reconciliation analysis allow for host switches as a process to explain incongruence (Page, 1994), and this convergence of the assumptions of reconciliation analysis and BPA may take us closer to a unified method for the study of host-parasite coevolution.

What are the general patterns of coevolution observed to date? There does not seem to be a common scenario, even among related host-parasite assemblages. Lice and their hosts, for instance, illustrate how patterns of coevolution are variable. The evolutionary history of several species of two related genera of chewing lice, and their hosts (members of the rodent family Geomyidae) appears to be one of rather strict cospeciation, with host switching playing a very minor role (Hafner and Nadler, 1988, 1990; Hafner and Page, 1995). Not only is there a clear congruence between the branching patterns of host and parasite phylogenies, but the timing of speciation events in both host and parasite lineages coincides remarkably well based on evidence from rates of molecular change. Not surprisingly, these louse species display a strict host specificity. Cospeciation also appears as the main explanatory process in the evolution of lice on seabirds (Paterson *et al.*, 1993; Paterson and Gray, 1996). Host-switching events are more important in other bird-louse associations (Clayton *et al.*, 1996), but cospeciation remains frequent. In contrast, host-switching events appear to have been very common, and cospeciation almost non-existent, in lice parasitic on several species of one genus of rock wallabies in Australia (Barker, 1991). To some extent the discrepancies between these studies may be due to differences of opinion among researchers, with the same evidence being interpreted differently by different people (Barker, 1994, 1996; Page *et al.*, 1996), but unidentified biological differences between the host-louse systems in these studies no doubt account for most of their contrasting results.

At present there have been too few studies on host-parasite coevolution to reach any verdict. One fact should not be forgotten, however: despite most parasite species being quite host specific, many are today found in more than one species of host. If we assume that the same parasite species cannot have more than one origin (i.e. it cannot evolve independently more than once), there are two ways in which low host specificity can evolve. Firstly, it can be the result of host switching, where a parasite repeatedly colonizes host lineages related or unrelated to that of its original host. Secondly, it can result from multiple speciation events in the host phylogeny without any corresponding speciation in the associated parasite lineage. This second possibility requires that gene flow be interrupted between parts of the host population but not within the parasite population. Both processes are equally possible and both require that the parasite be able to disperse efficiently.

The above discussion focused on the macroevolutionary history of host-parasite associations and host specificity. On a microevolutionary scale, many phenomena can facilitate host switching and subsequent decreases in host specificity. The physiological and ecological characteristics of hosts and parasites are obviously important and will be addressed in the next section. Independently of these, other processes can play a role. Hybridization between host species, for example, can create a genetic and ecological bridge between host species and allow the colonization of one host by parasites from the other. Le Brun *et al.* (1992) have documented such a situation among two related fish species. One of them, *Barbus meridionalis*, is host to the monogenean *Diplozoon gracile*, whereas the other, *B. barbus*, is never parasitized by *D. gracile* under natural conditions but is a perfectly suitable host in laboratory infections. The two congeneric fish species hybridize in nature, and the likelihood of

D. gracile infections among hybrids is proportional to the percentage of *B. meridionalis* genes in the hybrid (Figure 3.6). The microhabitat preferences of hybrids are also related to which of the parent species they most resemble, and determine the encounter rates between parasites and fish. It is easy to see how this could lead to a gradual extension of the parasite's spatial range to cover the microhabitats of both parent host species. Monogeneans generally form one of the most highly host-specific groups of parasites (see section 3.4). The above example illustrates that host switches can easily occur even in the most specific of parasites.

Just as parasites can be selected to increase the range of hosts in which they can successfully develop, they may also sometimes face selection for greater specialization through a narrowing of their range of suitable hosts. The growth and fecundity of any given parasite vary among host species (Poulin, 1996b). If selection can fine tune the mechanisms of host infection to ensure that fewer host species are encountered, then one would predict that host species in which development is suboptimal will be dropped. Often, though, there will be a trade-off between the ability to use many host species and the average fitness achieved in these hosts. Close adaptation to one host species can usually

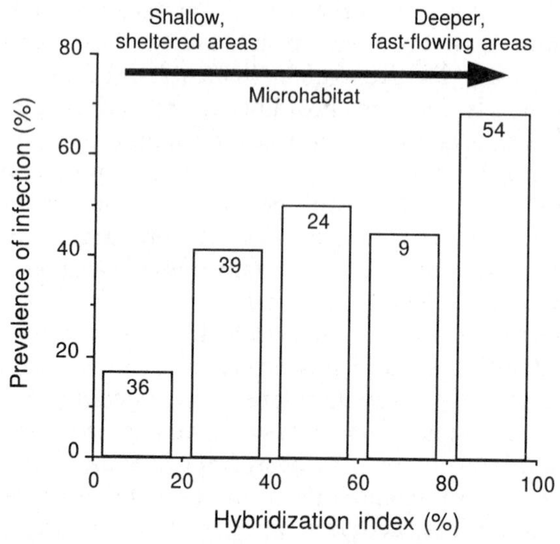

Figure 3.6 Prevalence of infection by the monogenean *Diplozoon gracilis* as a function of the degree of hybridization between the fish species *Barbus meridionalis* and *B. barbus*. A pure *B. barbus* fish has a score of 0% and a pure *B. meridionalis* specimen has a score of 100%; numbers of fish analysed in each hybridization class are given in the bars. Only *B. meridionalis* is a natural host of the parasite; as hybrids come genetically closer to *B. meridionalis*, they are found in microhabitats preferred by *B. meridionalis* and are more likely to harbour parasites. (Data from Le Brun *et al.*, 1992.)

be achieved only at the expense of adaptations to other host species. These trade-offs are often used to explain the host specificity of phytophagous arthropods (Fry, 1990).

The selection of greater host specificity in parasites can perhaps be seen on a finer scale, when comparing the specificity of different populations of the same parasite species. In some trematode species, for instance, there is evidence of adaptation to local host populations. In cross-infection experiments, miracidia of *Microphallus* sp. are more successful at infecting snails from the local snail population than allopatric snails (Lively, 1989), despite relatively high rates of gene flow among parasite populations (Dybdahl and Lively, 1996). Cercariae of *Diplostomum phoxini* are more successful at infecting local fish than fish of the same species but from allopatric populations (Ballabeni and Ward, 1993). Exceptions to this pattern are known and expected if parasite genotypes track local host genotypes with a lag (Morand *et al.*, 1996b). In general, though, after generations of isolation from other host genotypes, parasites may lose the ability to infect allopatric hosts in favour of a greater specialization for the local host genotypes. Alternatively, parasites can retain the ability to infect allopatric genotypes but achieve lower fitness when exploiting them. For example, a microsporidian protozoan parasite of the crustacean *Daphnia magna* achieves much higher levels of spore production on local hosts than on hosts from distant populations (Ebert, 1994).

A similar process may operate at the species level, with parasite species gradually losing the ability to exploit seldom-encountered host species and eventually excluding them altogether from their range of suitable alternatives. For instance, a strain of the parasitic nematode *Howardula aoronymphium*, a parasite of several *Drosophila* species in nature, apparently lost the ability to infect other species after being cultivated in the laboratory using a single host species (Jaenike, 1993). After only 3 years (or 50 parasite generations) in the laboratory culture, this strain became highly specific for the only host species that had been available in its recent evolutionary history. Two other strains of the nematode, maintained in the laboratory for the same time, did not lose the ability to infect natural hosts, suggesting that there is genetic variation for host specificity in this parasite, a prerequisite for the evolution of higher or lower specificity. Recent genetic evidence, however, suggests another explanation for the apparent increase in host specificity in *H. aoronymphum* (Jaenike, 1996a). There may have been two types of nematode in the initial sample used to start the culture: a highly specific type that survived, and a less-specific type that was lost over time.

The copepod *Lepeophtheirus europaensis*, parasitic on two species of flatfish with different habitats in the Mediterranean Sea, provides another example (De Meeüs *et al.*, 1992). It infects flounder in brackish lagoons and brill in marine habitats. In what is believed to be a speciation event in progress, individual parasites taken from the two different host species show genetic differences as well as physiological adaptations to the salinity of their hosts' respective habitats. This is a case where an ancestral parasite species capable of exploiting two host species gives rise to two species each specific to a single host species. In addition, *L. europaensis* populations infecting brill live in sympatry with *L. thompsoni*, a parasite of turbot. In the laboratory, the two congeneric species can meet and mate on turbot and produce viable hybrids, but in nature strict host specificity maintains the two species as distinct genetic entities (De Meeüs *et al.*, 1990, 1995).

Several important parasites of humans also appear to have expanded their host range by adding humans to their spectrum of suitable hosts, only later to abandon their initial vertebrate hosts to specialize on the newly colonized human hosts (Combes, 1990; Waters *et al.*, 1991). These findings indicate that the evolution of host specificity is not necessarily a sequence of host colonizations but may proceed in both directions, toward higher or lower specificity. Host specificity is just another continuous variable on which selection acts in no fixed direction; although greater specialization on fewer host species can be advantageous, it also links the fate of parasites to that of their hosts and can make highly host-specific parasites more prone to local extinction.

Before moving on to the conditions that may predispose certain parasites for low or high host specificity, the above ideas can be extended one step further. If parasites can become specialized for the local host population, why stop there? Why not specialize for the individual host itself? This is obviously not an option for the majority of parasites which do not spend more than one generation on the same host individual. The host being only an ephemeral habitat, the parasite's offspring must disperse and locate new hosts; they must be adapted to infect hosts of one or more particular species and not only the genotype of the host they just left. However, some rapidly dividing protozoan parasites do spend several generations inside one individual host after the host acquires an initial infection. Recent evidence suggests that over time, selection acts on the many dividing parasites inside the host and favours the parasite genotypes best suited for the particular conditions in that individual host (Seed and Sechelski, 1996). A very similar phenomenon occurs in phytophagous arthropods, which can complete thousands of generations on a single host plant (Karban, 1989). Clearly, parasites must retain the ability to colonize new host individuals and cannot become specialized exclusively for a certain host genotype. Rapidly dividing parasites, however, provide an example of a microevolutionary trend toward ultimate specificity.

3.3 DETERMINANTS OF HOST SPECIFICITY

To illustrate the constraints and pressures acting on the evolution of host specificity, I will use Combes' (1991b) filter concept. Two filters determine how many animals can be used as hosts by a parasite: an encounter filter and a compatibility filter (Figure 3.7). The encounter filter excludes all animals that the parasite cannot meet for ethological or ecological reasons. The compatibility filter eliminates species in which the parasite cannot survive and develop, for morphological, physiological or immunological reasons. Selection will act on these filters to increase or decrease their permeability and specificity; with respect to Figure 3.7, selection acts to change the diameter of the encounter filter or the angle of the compatibility filter. Other frameworks developed to illustrate the evolution of host specificity, such as mathematical models (Garnick, 1992a), lack the generality and simplicity of the filter concept. Here some of the factors that may influence the filters are discussed, and simple predictions suggested that can be tested with comparative data.

Properties of the parasites can determine what range of potential hosts will be encountered (Price, 1980). Selection has provided parasites with detailed genetic instructions to help them find a suitable host at each stage of the life cycle. Nevertheless, mistakes happen and are usually fatal. Several events can change the frequency of such mistakes through evolutionary time. For example, a new species with ecological attributes (diet, size, microhabitat preferences, etc.) identical to those of a parasite's

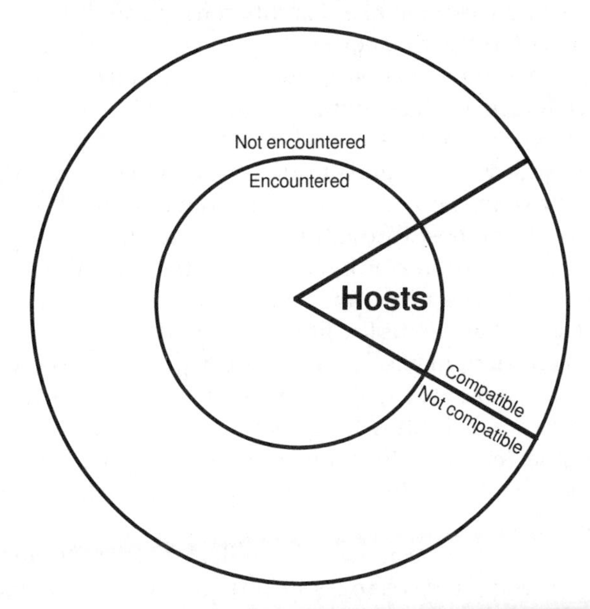

Figure 3.7 Schematic representation of the forces determining the range of host species used by a parasite. Of all potential host species (outer circle), only a subset are actually encountered by the parasite: the encounter filter is represented by the smaller inner circle. Another filter, the compatibility filter represented by the triangular area, excludes all species in which the parasite cannot develop. Some compatible species are never encountered, and *vice versa*. Only species passing through both filters can become hosts. (Modified from Combes, 1991b.)

only host, can invade the area occupied by the parasite and become more abundant than the host. Because suitable hosts would now become diluted in a sea of unsuitable targets, mistakes could become more common. Selection would favour any mutant parasite capable of surviving in and exploiting the invading species. The fitness of such parasites would not initially have to be equal to what could be achieved in the traditional host, it would just need to be greater than zero. Even if alternate targets are readily available, the likelihood of making mistakes during transmission and the opportunity for colonization of new hosts will depend on the mode of transmission.

The infective larvae of ectoparasites such as monogeneans and copepods are equipped to locate and infect the right species of host. Similarly, the mobile infective stages of many endoparasites must locate, attach to and penetrate the skin of their host, and possess several mechanisms for host location and recognition (see section 2.4). Among related species, the behaviour (Snyder and Janovy, 1996) or dispersal ability (Downes, 1989) of mobile infective stages can explain differences in host specificity, but these stages are typically equipped to encounter few host species. Other parasites can be transmitted only *via* direct physical contact or during copulation between their host individual and another individual; as physical contact between members of different species is unusual, opportunities for transferring to another host species are limited. Parasites using all the above transmission routes may therefore be less subject to mistakes than infective stages entering the host by ingestion. This latter route of infection relies on hosts eating infective stages or infected intermediate hosts and does not prevent non-host species from ingesting the parasites (Noble *et al.*, 1989). All else being equal, we may predict (i) that parasites entering the host through the mouth are less host specific than parasites using other routes of infection; in other words, the encounter filter (Figure 3.7) allows more host species to reach parasites using the oral route of infection. Complex life cycles, with their many steps and modes of transmission, may select for flexibility in the parasite's range of acceptable hosts. Noble *et al.* (1989) suggested that (ii) parasites with complex life cycles should be less host specific than parasites with simple, direct cycles; and (iii) parasites with complex life cycles should be more specific in their choice of intermediate hosts than when settling on a definitive host.

Characteristics of the hosts will also influence opportunities for host switches and evolutionary changes in host specificity. Suitable but not-yet-colonized hosts are those that provide parasites with suitable living conditions and opportunities for the completion of the life cycle. They must therefore be similar to the hosts used by a parasite. The similarity can be the result of phylogenetic inheritance, if the suitable host is related to the parasite's current host, or convergence, if the suitable host has evolved independently to resemble the parasite's current host (Kennedy, 1975). Obviously, in most cases similar hosts will be relatives, so that we can predict that (iv) parasites of hosts belonging to species-rich taxa will be less host specific than parasites of hosts with no or few relatives, because of a difference in the availability of suitable hosts. Exploitation of a host from a speciose taxon means that the compatibility filter is wide open (Figure 3.7).

Other host traits may also affect the evolution of host specificity. For instance, it is widely recognized that a high degree of host specificity can lead to a high risk of local extinction when host abundance drops below a critical level and the parasite is unable to use other available animals (Kennedy, 1975). Combes (1995) suggested that parasites exploiting hosts with unstable populations may be selected to maintain a wide host spectrum. The ability to exploit a wide range of hosts comes at a cost – a lower degree of specialization can mean lower fitness in general. Because of this cost, perhaps a better long-term solution against the risk of extinction would be to maintain a wide-open encounter filter but a very exclusive compatibility filter (see Figure 3.7 and Combes, 1991b). Similarly, parasite populations on islands may be selected to exploit a wider range of host species because of the greater risk of extinction faced by host species in insular habitats (Freeland, 1983). Interesting as this may be, such considerations do not lead to any easily testable predictions. The next section will demonstrate how the available evidence supports the few predictions made above.

3.4 OBSERVED PATTERNS OF HOST SPECIFICITY

In any parasite taxa, the majority of parasite species tend to be very host specific. Animals in general tend to be limited in their distribution among potential habitats. For instance, the sizes of the geographical ranges of free-living species within any given taxon vary greatly, but their distribution is typically strongly right-skewed. However, the observed distributions are usually log normal, i.e. when geographical range sizes are log transformed their distribution becomes normal (Gaston, 1996). The predominance of small geographical ranges is an artefact of scale, and the skewness is not pronounced enough to survive the normalizing effect of a log transformation. Distributions of host specificities of parasite species within a taxon, on the other hand, are strongly right-skewed before and even after log transformation (Figure 3.8). This indicates a very strong tendency for parasites to be highly host specific. Low host specificity is very much the exception.

Despite this tendency, there exists variability in host specificity between related parasite taxa. Can some of it be explained by differences in the route of transmission? Finding related parasite taxa that differ only

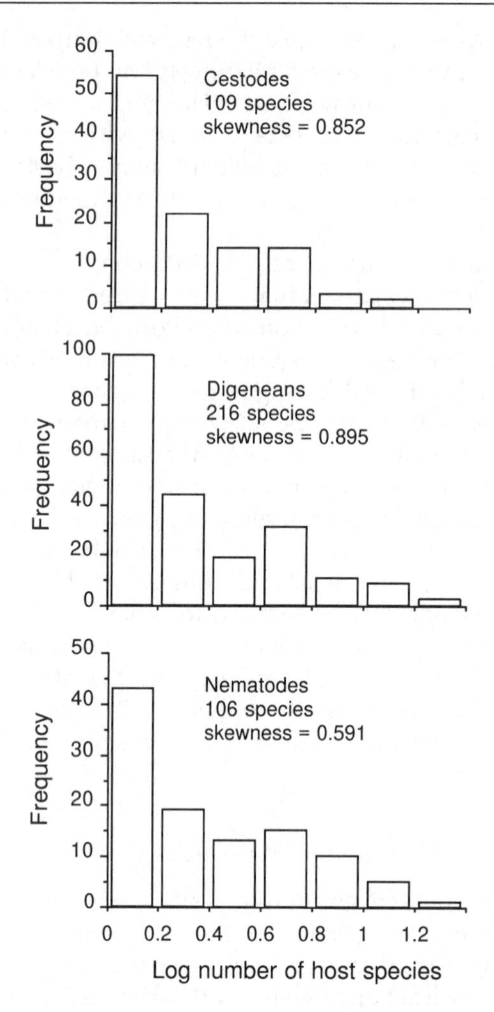

Figure 3.8 Frequency distribution of the number of known host species for cestodes, digeneans and nematodes parasitic in birds. The index of skewness presented is the g_1 of Sokal and Rohlf (1981). Even though the distributions use log-transformed numbers of host species, all three are significantly right-skewed and are not lognormal. (Data from Gregory *et al*., 1991.)

in how they infect their host is not easy. Perhaps the best group for such a contrast would be digeneans in which one lineage penetrates the vertebrate definitive host through the skin, whereas other digeneans reach it *via* ingestion. Detailed information is not available to perform a robust comparison, but skin-penetrating schistosomes often have a wide repertoire of definitive hosts (Noble *et al.*, 1989), whereas many surveys report that among other digeneans close to 50% of species occur in a single host species (Gregory *et al.*, 1991; Poulin, 1992a). Thus the prediction that orally infecting parasites are less specific than those penetrating through the skin may not be supported.

Monogeneans are widely regarded as highly host specific; typically, in any data set, much more than half of monogenean species are known from a single host species (Kennedy, 1975; Poulin, 1992a; Rohde, 1993). They are capable of evolutionary host switches (Guégan and Agnèse, 1991) but generally are so specific that their mere presence is often enough to identify the host species (Lambert and El Gharbi, 1995). Why are they much more host specific than their relatives, the digeneans and cestodes? Monogeneans attach to the skin of their aquatic vertebrate host rather than infecting it orally, but also differ from digeneans and cestodes in having simple, direct life cycles. Their route of infection, the simplicity of their life cycle, or some other characteristic such as the specialization of their attachment organs, may all have contributed to their specificity. There is no way to tell at present which, if any, of these factors affected the evolution of host specificity in monogeneans.

Comparisons between a large number of sexually transmitted parasites and their relatives that use a different route of infection have indicated that sexually transmitted parasites utilize a much narrower range of host species (Lockhart *et al.*, 1996). Given that sexual contact between hosts is almost exclusively intraspecific, this is not surprising. There is a need for more comparative analyses of this nature, which attempt to control for phylogenetic influences in order to determine if the mode of transmission itself selects for a certain level of host specificity.

There has been only one true test of the prediction that parasites with complex life cycles are less host specific than their relatives with direct life cycles. The average number of known vertebrate host species of 35 ascaridoid nematode species with simple life cycles was found not to differ from that of 334 ascaridoid species with complex life cycles (Morand, 1996a). In this comparison, parasite species were treated as independent from one another regardless of their taxonomy and thus phylogenetic influences were not removed. Still, this result provides no support for the prediction that host specificity is relaxed in parasites evolving complex life cycles. Yet another prediction made in the previous section, that parasites with complex life cycles should display a higher specificity for their intermediate host than for their definitive host, remains mostly untested. The specificity of some helminths with complex life cycles for their intermediate host can prove lower than expected when detailed experimental infections are performed (e.g. Dupont and Gabrion, 1987). Comparisons from a wide range of parasite species are needed before the prediction can be evaluated adequately, however.

The expansion of the number of host species used by the process of host switching should be facilitated when alternative species physiologically and ecologically similar to the host are available. Among parasites of Canadian freshwater fish, this certainly appears to be the case (Figure 3.9) – parasites exploiting hosts belonging to species-rich taxa have colonized several other potential hosts closely related to the original host (Poulin, 1992a). The same apparently occurred among digeneans exploiting coral reef fish on the Great Barrier Reef (Barker *et al.*, 1994). The average number of host species used by 18 digenean species found among 39 species of pomacentrid fish was almost five, with four parasite species each occurring in more than 10 host species. In addition, nine of the 18 digenean species were also found in hosts from other fish families. The large number of related fish species, living in sympatry at high densities, probably made host switching easy. Host switches between unrelated hosts can also take place, but probably under unusual circumstances. For instance, the colonizations of humans by parasites from unrelated hosts (e.g. schistosomes or the nematode *Onchocerca*; see discussion in Combes, 1995) are probably consequences of the rapid geographical expansion of humans, their modified behaviour and their impact on the environment.

There is a great gap in our understanding of the evolution of host specificity. Few solid patterns have been reported, probably not because they do not exist but because no one has bothered looking for them.

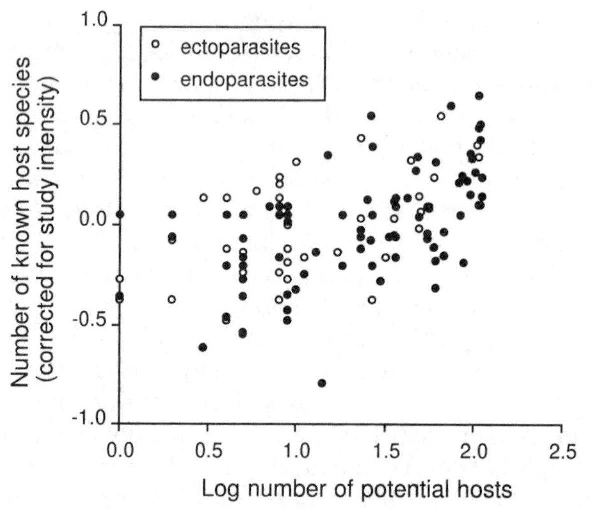

Figure 3.9 Relationship between the number of known host species and the number of potential hosts available among metazoan genera of ectoparasites ($N = 41$, $r = 0.514$, $P < 0.001$) and endoparasites ($N = 90$, $r = 0.511$, $P < 0.001$) of Canadian freshwater fish. The number of known hosts was corrected for study intensity; the scores used are the residuals of the regression of log number of known hosts on log number of published records (see Figure 3.1). The number of potential hosts is obtained by summing up all fish species belonging to the family or families that include the parasite's known hosts. (Data from Poulin, 1997a.)

Some of the few known patterns are yet to receive a satisfactory explanation. For instance, there is a latitudinal gradient in the host specificity of digeneans parasitic in marine fish; digenean species from the tropics are restricted to fewer host species than their relatives in colder seas (Rohde, 1993). Monogeneans, on the other hand, display consistently high host specificity from low to high latitudes. Why the two taxa follow different trends is not clear, and this exemplifies how little we know about the evolution of host-parasite interactions and host specificity.

3.5 SUMMARY

One of the most fundamental characteristics of a parasite is the spectrum of host species used at each stage in its life cycle. This information is not available for many parasite species; for others, information is available but not yet understood. Two avenues are open for the investigation of the evolutionary forces shaping host specificity. Firstly, recent and future developments in cladistics and the reconstruction of parasite phylogenies should provide an increasing number of historical accounts of associations with hosts. Refinements in existing methods of reconciling host and parasite phylogenies may allow us to infer extinction or host-switching events and make predictions about the conditions favouring the colonization of new host lineages. The second approach should be to test these predictions with comparative analyses of the existing data in many different groups of parasites. As comparative biology gains in popularity among parasitologists, the next few years should see great strides toward elucidating the many aspects of host-parasite associations.

Evolution of parasite life history strategies 4

Parasites must complete difficult journeys through several hosts as well as through the external environment. Evolution has sometimes favoured longer and more complex journeys. Sometimes it has opened up alternative paths through the same cycle by relaxing host specificity. The length, complexity and flexibility of the life cycle are the outcome of selection acting on the cycle as a unit of selection. But evolution has also shaped the organism itself, and this chapter now focuses on the parasite as an organism rather than as a life cycle.

Several laws or rules have been proposed to describe the evolution of parasitic organisms, most of which have not survived the test of time and robust analysis (Brooks and McLennan, 1993a; Poulin, 1995d). Two of the most pervasive misconceptions about parasite evolution are that parasites evolve toward reduced body size and toward higher fecundity. Parasites are smaller than their hosts; that is a fact, but an irrelevant one when considering the evolution of parasite body size. The interesting comparison is not between the size of the parasite and that of its host, but between the size of the parasite and that of either its free-living ancestor or its closest free-living relative. Similarly, parasite fecundity is often pitted against that of free-living organisms. Many textbooks dazzle their readers with tables listing the egg output of some of the most fecund parasites (e.g. Esch and Fernández, 1993). Again, the truly interesting questions relate to how these numbers compare to egg production in the parasites' free-living ancestors or closest free-living relatives, and why some parasites exhibit less than remarkable fecundity. Taxa that contain both parasitic and free-living members, such as crustaceans and nematodes, are very promising for such investigations.

Life history strategies are combinations of demographic traits such as body size, life span, age at maturity, fecundity and offspring size, that have been favoured by selection because they result in higher fitness in particular environments than other possible combinations. The simultaneous maximization of several traits may be impossible because of phylogenetic, physiological or physical constraints. Thus, the combinations of life history traits that can evolve are limited, and are characterized by trade-offs between various pairs of traits (Stearns, 1989). For instance, any investment in offspring size may come at the expense of fecundity. External pressures determine which way particular trade-offs will go, and the emerging set of life history traits is expected to show some association with the organism's mode of life or habitat (Partridge and Harvey, 1988; Southwood, 1988). Other reviews give detailed explanations of life history theory (Roff, 1992; Stearns, 1992) or how it applies to parasitic organisms in general (Poulin, 1996b). This chapter explores the evolution of life history strategies in parasites, but with special emphasis on body size and egg production.

4.1 PHENOTYPIC PLASTICITY AND ADAPTATION

In populations of parasites or free-living organisms, adjustments of life-history strategies to changing environmental conditions can take place through two distinct but mutually compatible processes (Stearns, 1992). Firstly, a given genotype may be capable of producing a variety of phenotypes under different conditions. In animals living in habitats that are highly variable spatially or over short time periods, a flexible developmental schedule would be highly advantageous. The set of possible phenotypes (referred to as a reaction norm) offered by a single genotype allows an organism the possibility of adjusting to local and current conditions through small changes in development, resulting in the life-history strategy best suited to those conditions. Phenotypic plasticity allows immediate responses to environmental changes and does not result in changes in genotypic frequencies in the population.

Secondly, if environmental changes persist in time, a true evolutionary response can occur, involving a shift in gene frequencies in the population. The genotypes producing sets of phenotypes better suited to the new conditions will be favoured, and can spread through the population over time. In contrast to phenotypic plasticity, adaptive genetic responses can take place only over several generations and not within a single generation. The two processes, however, are compatible and can operate jointly. For instance, if new environmental conditions favour large body sizes, individual organisms in one generation can opt to grow to the upper limit set by their genotype, while selection acting over several generations can increase the mean or upper limit of sizes that organisms can reach.

Parasites often display considerable phenotypic plasticity in life-history traits. For example, in *Triaenophorus crassus*, a cestode parasite of freshwater fish, mature and gravid individuals of the same age range in mass between 5.7 and 124 mg (Shostak and Dick, 1987). The largest adult individual in the population is over 20 times larger than the smallest. In the nematode *Raphidascaris acus*, also a parasite of freshwater fish, gravid female worms range in mass between 0.7 and 61.2 mg, the largest being almost 90 times the size of the smallest (Szalai and Dick, 1989). The ranges of fecundity values among individuals of these two species are even greater. The relative scale of the phenotypic plasticity displayed by these worms greatly exceeds that shown by free-living organisms. Such large differences in the adult size or fecundity of conspecific parasites from the same population are the result of the distribution of parasites among their hosts (see Chapter 6) – the aggregation of many parasites in a few hosts means that many worms live in crowded conditions and may not be able to reach their maximum adult size. Parasites that do not share their host with conspecifics, on the other hand, are able to reach their maximum potential size. Thus growing conditions in the host are unpredictable, and not all parasites in a population live in the same conditions. A flexible developmental programme therefore affords parasites the opportunity to make the best of difficult situations: they grow as close as they can to their maximum potential size, but can still mature at small sizes if they have no other choice.

Clearly, differences between individuals in a population are not entirely the product of phenotypic plasticity. Different individuals have different genotypes, and the source of differences in phenotypes can also have a genetic basis. Many variables are associated with the expression of life-history traits in individual parasites (Poulin, 1996b). Because adaptive responses are not easily distinguished from phenotypic plasticity at the level of the individual organism, the remainder of this chapter examines adaptive adjustments of life-history strategies using comparisons among species or higher taxa. At these levels, differences are surely the product of adaptive genetic changes. To compare species, however, one must use species-typical values for traits such as body size or fecundity. If these traits are as plastic within species as in the two examples given above, how can we be sure that the values used are indeed representative of the whole species? The answer is that we cannot; one can only hope that averages provide good estimates of a species' strategy. Also, all studies to date have investigated average species values and none have examined the variance in life-history traits. Variance of a trait may not only constrain evolutionary rates, but may also itself be subject to selection. Different parasite taxa may exhibit different levels of variability in life-history traits as a result of selection in hosts or habitats that differ in predictability, and exploring how this variability evolved would be particularly interesting.

4.2 PARASITE BODY SIZE

Most animal lineages undergo increases in body size over evolutionary time, a phenomenon known as Cope's Rule and supported by some fossil evidence (see Stanley, 1973; Bonner, 1988; Jablonski, 1996). In contrast, lineages making an evolutionary transition from a free-living existence to a parasitic mode of life are often assumed to evolve in the opposite direction, toward smaller body sizes (Price, 1980; Hanken and Wake, 1993). The physically restricted habitats of parasites are believed to have been the key factor promoting a reduction in their body size. The unstated and untested assumption behind this line of reasoning is that the free-living ancestors of modern parasites had larger body sizes than their parasitic descendants. Parasitism had multiple origins, and precursors of parasites came in all sizes and shapes. Evolution of parasite body size has no doubt proceeded differently in different lineages. This section begins with a comparison between the body sizes of parasites and their free-living relatives, and attempts to reconstruct the evolution of parasite body size in order to determine the direction of change following the adoption of parasitism as a lifestyle. The discussion then moves on to examine what other factors, related either to the host or to the environment, have played a role in the evolution of parasite body size. Finally, the forces creating a dimorphism in size between males and females in dioecious parasites will be explored.

4.2.1 CHANGES IN SIZE AS ADAPTATIONS TO PARASITISM

Within large taxa comprising both free-living and parasitic groups, quantitative comparisons between the sizes of these groups indicate clearly that parasites are as large as or larger than their free-living counterparts (Figures 4.1 and 4.2). There should therefore be no *a priori* reason to believe that parasites evolve to be smaller than free-living animals.

One way of contrasting the evolution of body size in parasites and their free-living relatives is to compare the shape of the frequency distributions of their body sizes. The shape of a body-size distribution can indicate which size classes have undergone extensive diversification and which remain species-poor, and can thus reflect the selection acting on body size within a large taxon. Almost universally, body-size distributions are right-skewed, even on a log scale, such that the small-bodied classes invariably include much more species than the large-bodied ones (Blackburn and Gaston, 1994a). Many taxa of parasites, however, depart from the typical right-skew pattern (Poulin and Morand, 1997). Log-normal distributions are common among ectoparasitic taxa such as monogeneans, copepods and isopods parasitic on fish. In these parasites, large size leads to higher fecundity but also to a greater likelihood of being dislodged from the external surfaces of the host, and selection may favour parasites of intermediate size. Ticks, however, show the usual right-skewed distribution of body sizes. Hosts of ticks, unlike fish, are capable of self-grooming, and this selective pressure may have pushed tick body sizes toward the smaller end

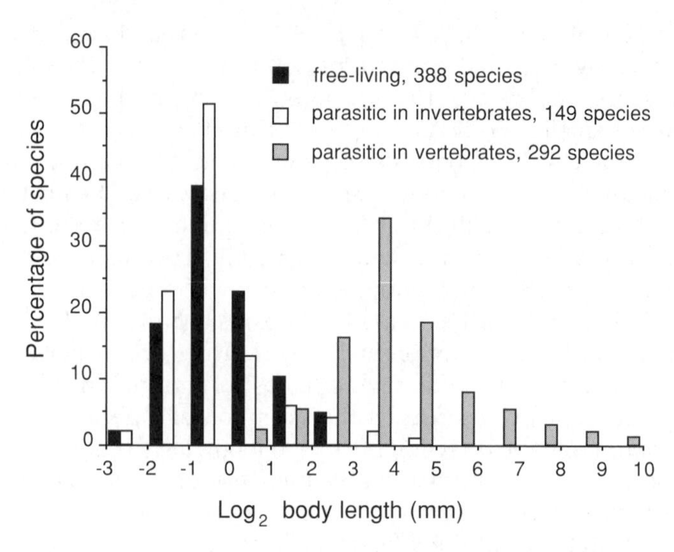

Figure 4.1 Frequency distribution of nematode body sizes with respect to mode of life and type of host. (Modified from Kirchner *et al.*, 1980.)

of their spectrum (Poulin and Morand, 1997). Internal parasites for which data were available, i.e. nematodes and digeneans, show right-skewed distributions of body sizes. The microhabitats available to these parasites may be physically constrained such that more niches are available for small-bodied endoparasites than for large ones. This would lead to higher speciation rates among small-bodied parasite taxa or to reduction in body size in the large-bodied taxa. Examinations of body-size distributions therefore suggest that evolution of body sizes proceed in different directions in different parasitic groups, and not always toward a reduction in size.

There are two problems with the use of body-size distributions to infer evolutionary trends. First, if the diversification of taxa within the same size class is independent of body size, treating species as independent observations introduces a bias in the analysis. Among monogenean species, for instance, the majority of species are in the smallest size class (Figure 4.3). When the body size of genera are examined instead, this trend weakens considerably. Some small-bodied genera, e.g. *Dactylogyrus* and *Gyrodactylus*, have diversified greatly but others have not (Poulin, 1996c), and therefore body size itself is not a correlate of evolutionary success. Using species body-size distributions may thus lead to results confounded by phylogenetic influences.

Secondly, the patterns observed may be artefacts of our incomplete knowledge of certain taxa. We can examine the body-size distributions only of known species, not of all species. Many small species go unnoticed and therefore undescribed for long periods; as they are found and added

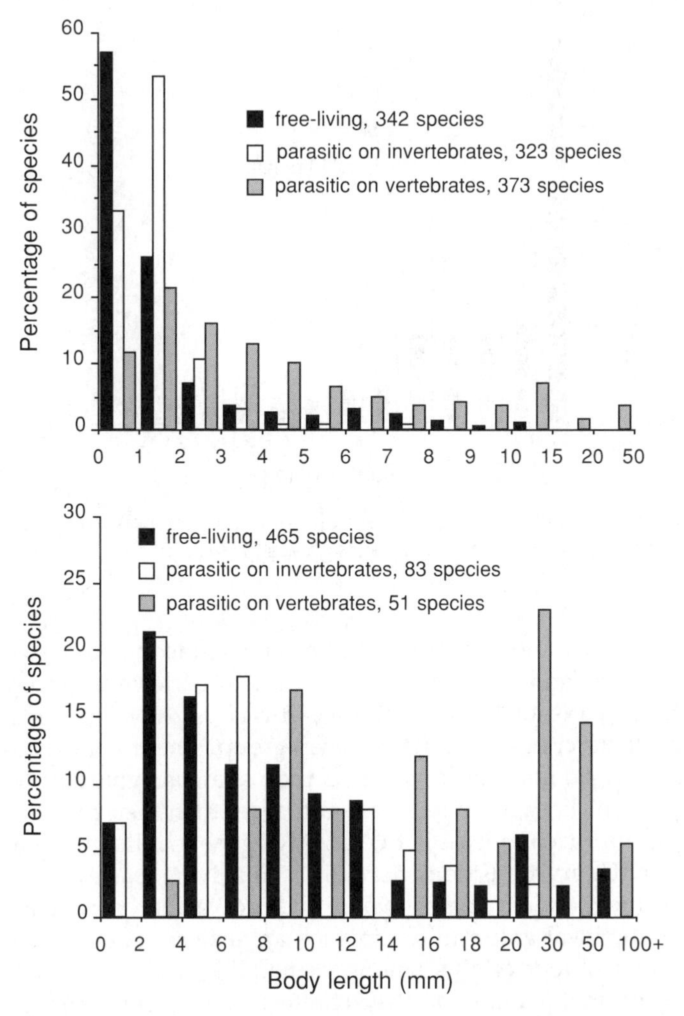

Figure 4.2 Frequency distribution of copepod (above) and isopod (below) body sizes with respect to mode of life and type of host. Terrestrial isopods are excluded, and only isopods that are obligate parasites of fish as adults (i.e. family Cymothoidae) are included in the vertebrate parasite category. Note contraction of scale on right-hand side of figure. (Data from Poulin, 1995b, c.)

to the body-size distribution of known species, the distribution becomes increasingly right-skewed (Blackburn and Gaston, 1994b). In some groups of parasites but not others, there is a negative relationship between species body size and date of description (Poulin, 1996a). We may be left only with the smallest species to find and describe in some taxa, whereas much of the diversity of other taxa may have escaped us thus far.

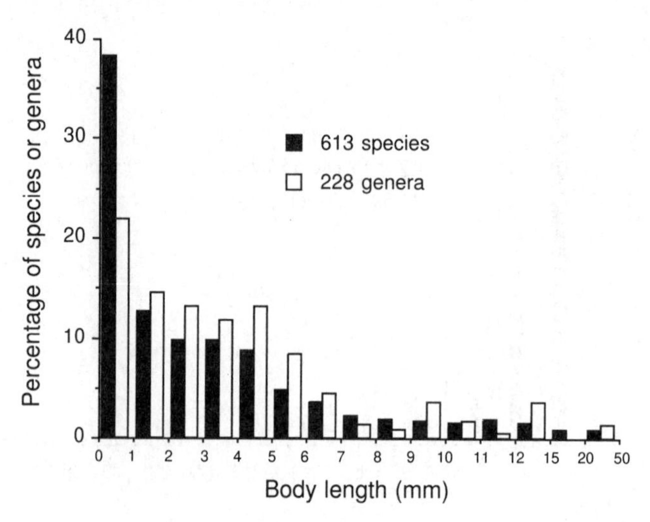

Figure 4.3 Frequency distribution of body sizes among monogenean species and genera. Note contraction of scale on right-hand side of figure. (From Poulin, 1996c.)

Another, much more rigorous method of contrasting body-size evolution in parasitic and free-living taxa consists of performing several independent comparisons between the mean body size of a parasitic taxon and that of its closest free-living sister group. These comparisons can then be analysed statistically to see if they show a significant tendency for the parasitic lineages to be smaller or larger than their free-living relatives. This approach at least is free of phylogenetic influences because it treats evolutionary transitions between a free-living existence and parasitism as independent observations and not actual taxa. Nematodes would be the ideal group to use in such an analysis. Parasitism has had several independent origins among nematodes, and different types of hosts (plants, invertebrates and vertebrates) have been colonized on several occasions. Unfortunately, a robust and comprehensive phylogeny of nematodes is required to derive the proper contrasts between sister groups differing in mode of life, and none is currently available.

To date, this approach has been used to examine the evolution of body size only in parasitic crustaceans. In the phylogenetic history of copepods, parasitism on invertebrate hosts evolved on nine separate occasions from free-living ancestors (Poulin, 1995b). Generally, the parasitic branch shows a larger body size than its sister branch which remained free-living; however, the average difference is not large enough to be significant. Parasitism of fish never evolved directly from a free-living ancestor among copepods; but on nine separate occasions a lineage parasitic on invertebrates made the transition to parasitism on fish. Contrasts between these nine pairs of sister branches indicate that these transitions from invertebrates to fish were followed by significant increases in body size, the average greater than 3 mm in body length being huge for copepods (Poulin, 1995b). These trends are apparent in the distribution of copepod body sizes (see Figure 4.2). The history of copepod parasites is thus one of increases in body size.

Body-size distributions of isopods also suggest little difference between free-living isopods and isopods parasitic on invertebrates, with fish-parasitic isopods being considerably larger (Figure 4.2). This is a deceptive picture, however. There were few transitions to parasitism in the phylogeny of isopods, but in all cases the branch that became parasitic displayed a smaller body size than its free-living sister branch (Poulin, 1995c). This counter-intuitive result can be reconciled with Figure 4.2. The large-bodied Cymothoidae, parasitic on fish, evolved from morphologically similar ancestors that were only facultative parasites of fish. These in turn had evolved from free-living ancestors, whose extant descendants are the Anuropidae, a family of giant isopods inhabiting oceanic depths. Thus while cymothoids are larger than most free-living isopods, they are most closely related to some of the largest of these free-living taxa, and have apparently evolved toward slightly smaller bodies than those of their ancestors.

Finally, the phylogenetic history of amphipods suggests 17 independent transitions from a free-living existence to an obligate association with invertebrates. Although amphipod symbionts of invertebrates may not be true parasites, they face the same transmission problems and other pressures as those acting on parasites, and trends in the evolution of their body size may be instructive. In the majority of cases (13 out of 17), the amphipod lineage switching to symbiosis on invertebrates shows a smaller body size than its free-living sister group (Poulin and Hamilton, 1995). The average decrease in body size for all 17 contrasts, though, was small and not significant.

Parasitism leads to increases in body size in copepods, but to decreases in isopods and amphipods. These groups illustrate how variable the evolution of parasite body size can be, and how much it depends on the size of the ancestor. There are no rules governing body-size evolution, and each lineage may proceed down a different path. Similar studies on nematodes and perhaps other helminths would serve to extend this conclusion to all parasites.

4.2.2 CORRELATES OF BODY SIZE

Following a transition to parasitism in a branch of a phylogenetic tree, and whatever general direction is taken by evolution with respect to body size, the lineage will undergo diversification and give rise to many species that do not all have identical body size. Various factors can act to shape body size, by placing pressures or constraints on its evolution (Poulin, 1996b). These factors can originate from the host or from the external environment.

Host body size is one of the most obvious host-related factors that can affect the evolution of parasite body size. There are often differences in size between parasites of invertebrates and their relatives exploiting the larger vertebrates (see Figures 4.1 and 4.2). Within any host taxon, such as mammals, space constraints faced by parasites may vary inversely with host body size. The volume of the lumen in the mammalian gut increases as we go from a mouse to an elephant. Large size is correlated with fecundity in most animals, including parasites (Peters, 1983); given the space, selection should favour larger-bodied parasites over their smaller conspecifics, and drive the evolution of body size toward larger sizes. Larger hosts may also provide a greater supply of nutrients to parasites. As life span correlates positively with body size, larger hosts are also likely to provide parasites with a more permanent habitat, which can favour individual parasites that delay maturity and reach larger sizes (Stearns, 1992). These are some of the reasons why we may expect a relationship between host size and parasite body size.

One fact is often overlooked, however. Whereas large parasites can live only in large-bodied hosts, small parasites can exploit a wide range of host sizes. The variability in parasite body size may therefore increase along with mean parasite body size as we move from small to large hosts. This would weaken the relationship between host size and parasite size, but would not eliminate it altogether if it exists.

In some parasite species, the body size of individual parasites in the population correlates with the body size of their host individual (see Poulin, 1996b). There have been a few attempts to find such a relationship across species or higher taxa. Kirk (1991) found a positive relationship between the body size of flea species and that of their bird and mammal hosts. Harvey and Keymer (1991) found that after controlling for phylogenetic influences, host size and parasite body size correlated positively among species of chewing lice infecting rodents and species of pinworms parasitizing primates. Extending the analysis to oxyurid nematodes parasitic of both invertebrates and vertebrates, Morand *et al.* (1996a) found a strong relationship between parasite body size and host size, independent of parasite or host phylogeny. A comparative analysis across digenean taxa gave contrasting results depending on the analytical method used to control for phylogenetic effects, but also provided evidence for a positive relationship between vertebrate host mass and digenean adult body size (Poulin, 1997b). Finally, host size also correlates positively with parasite size among species of rhizocephalans, a group of crustaceans parasitic in other crustaceans (Poulin and Hamilton, 1997).

In comparison, few studies failed to find an association between host size and parasite size (e.g. among copepods parasitic on fish; Poulin, 1995b). In some cases, the relationship was not apparent across genera or families when phylogenetic effects were removed, but became clear in analyses across species within genera, or genera within families, when taxa were treated as independent observations. For instance, a strong relationship between host size and parasite size across genera within the diverse monogenean family Dactylogyridae failed to translate into a similar relationship among monogenean families (Poulin, 1996c). Also, host size and parasite size are consistently and strongly correlated among species in the most speciose genera of ticks, but not across tick genera (Poulin, 1997c).

Given the preceding results, it is probably fair to say that large host size facilitates the evolution of parasites toward large body size and the high fecundity that goes with it. This can be mediated by other factors. For instance, larger hosts are longer-lived than small hosts, and place fewer temporal constraints on parasite growth (Harvey and Keymer, 1991; Sorci *et al.*, 1997). Various other forces will intervene to determine parasite body size, however. For example, schistosomes live in the blood vessels of their host and their site of infection will place an upper limit on the size they can attain, independently of the body size of the host. Some important factors may even come from the external environment, as environmental variables can shape life-history traits in free-living animals. Among free-living poikilothermic animals, for instance, temperature during development usually affects growth rates and body size, with most animals reaching smaller sizes at higher temperatures (Atkinson, 1994). This effect has led to adaptive adjustments in life-history traits in free-living animals, with species living in cold environments differing in body size and other traits from their relatives in warmer habitats (Sibly and Calow, 1986). Internal parasites are exposed to outside conditions only for brief periods, and most if not all growth takes place in the host. External factors such as temperature may still have some impact on them, for example if these factors influence the physiology of the host with cascading effects on parasites, or if the host does not buffer the parasite against external conditions. This latter scenario was not supported in a comparison of digenean taxa living in poikilothermic definitive hosts with their sister taxa living in homeothermic birds and mammals: host type had no detectable effect on digenean body size (Poulin, 1997b).

Ectoparasites, on the other hand, are exposed to external conditions throughout their entire life and could show patterns of body-size variation similar to those reported for free-living organisms. For instance, in parasitic crustaceans body size tends to increase with increases in both latitude and water depth, two environmental correlates of water temperature (Poulin, 1995b, c; Poulin and Hamilton, 1995). In comparisons between sister taxa, the lineage inhabiting higher latitudes or deeper water tends to be larger-bodied than its relatives in warmer waters, independently of other variables such as host type (Figure 4.4). A similar pattern is observed among monogenean taxa (Poulin, 1996c). In ticks, however, body size showed no correlation with average air temperature in analyses across species or higher taxa (Poulin, 1997c). Again, these examples illustrate the many different responses of parasite body size to selective pressures, and reinforce the point that the evolution of body size in parasites does not follow any given route.

4.2.3 SEXUAL SIZE DIMORPHISM IN PARASITES

The above discussion applied to parasite body size in general. In many dioecious parasites, however, another factor creates differences in body size within species. That factor is sex, and its effect is expressed as differences, sometimes very large, between the body sizes of males and females. The causes behind these differences vary according to mating systems, the dependence of fecundity on size, etc. (Shine, 1989). Whatever their nature, asymmetric selective pressures on males and females can result in different evolutionary responses.

Because fecundity is related to body size and because the transmission of most parasites involves a series of unlikely events, we may

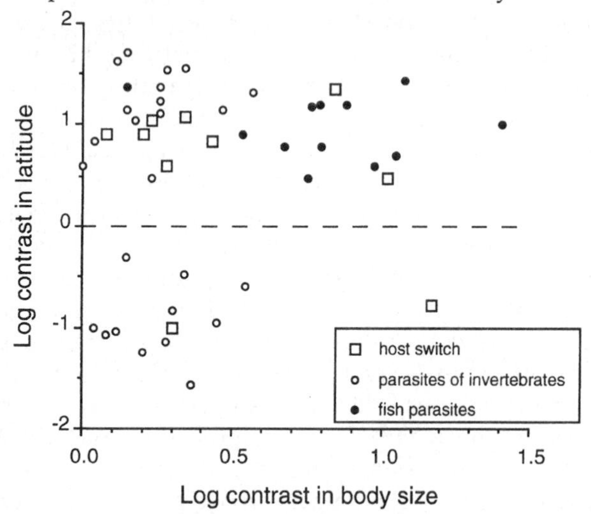

Figure 4.4 Relationship between latitude and body size in parasitic copepods. Each point represents a contrast between two sister branches in the copepod phylogeny. Values for the branch with smallest body size were subtracted from the branch with greater body size, giving contrasts in body size that are all positive. In most cases the contrast in latitude is also positive, indicating that copepod taxa with greater body size inhabit higher latitudes than their sister taxa. Host switches represent cases in which one taxon is parasitic on invertebrates and its sister taxon is parasitic on fish. (Modified from Poulin, 1995b.)

expect selection to exert more pressure on female body size than male body size. This appears to be the case in most parasitic crustaceans, where males are often dwarves compared to females (Raibaut and Trilles, 1993). In copepods, free-living lineages that switched to parasitism on invertebrates experienced small increases in both male and female body size, with little resulting change in sexual size dimorphism (Figure 4.5). However, when lineages parasitic on invertebrates switched to fish hosts, female body size increased substantially while male size did not change significantly. The result is a highly female-biased sexual size dimorphism among copepods parasitic on fish. A possible explanation for the influence of host type on sexual size dimorphism may be that fish are mobile and more difficult to infect than the generally sessile invertebrates serving as hosts to copepods. Selection could have favoured high fecundity, and thus large body size, in female copepods parasitic on fish to a greater extent than in their sister taxa parasitic on invertebrates (Poulin, 1995b, 1996d).

A female-biased size dimorphism also exists in oxyurid nematodes, but in contrast to copepods it is more pronounced in taxa exploiting invertebrates than in those parasitic in vertebrates (Morand and Hugot, 1997). Competition for food among nematodes of both sexes may be intense in the digestive tract of small invertebrate hosts. In addition, sexual selection may favour early maturation in male parasitic nematodes (Morand and Hugot, 1997). These factors may have acted in tandem to push the evolution of male oxyurids toward small body sizes, especially in parasites of invertebrates.

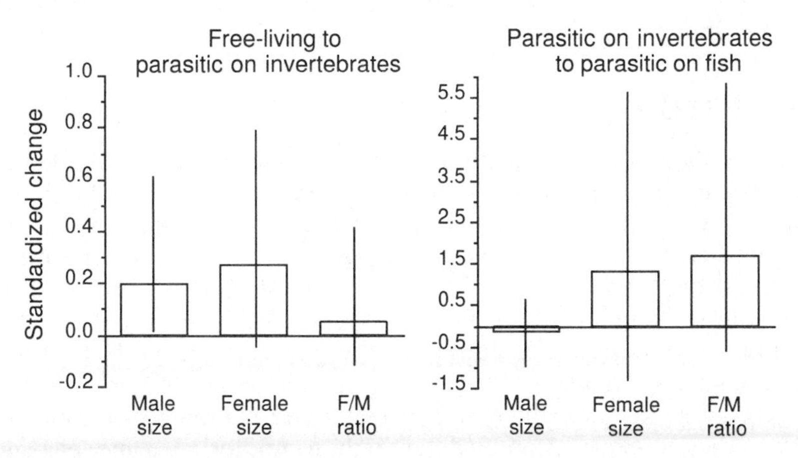

Figure 4.5 Mean (and range) change in body size and female-to-male size ratio for nine transitions from free-living to parasitic on invertebrates, and nine transitions from parasitic on invertebrates to parasitic on fish, during the phylogenetic history of copepods. Change estimated as the difference between sister taxa with different modes of life or different hosts, standardized for branch length. (Data from Poulin, 1996d.)

The situation may be different in some helminths. Schistosomes evolved from hermaphrodite ancestors and now have separate sexes, with males having larger body sizes than females. Females lodge themselves in a groove on the ventral surface of the male and the two form long-lasting monogamous pairings. One hypothesis explaining male-biased sexual size dimorphism in schistosomes is that a division of labour has been favoured by selection (Basch, 1990). Females have lost much of their body musculature and become highly specialized for reproduction. The large males provide all the muscle that the pair needs to move and feed. There is variability in sexual size dimorphism across schistosome genera, ranging from moderate (e.g. *Schistosomatium*) to extreme (e.g. *Schistosoma*). The division of labour hypothesis and other hypotheses could be tested in comparative analyses that take advantage of this variability among schistosome taxa.

The constraints imposed by phylogeny on sexual size dimorphism are apparent when two unrelated groups under similar selective pressures are contrasted with one another. Nematomorphs, which emerge from insect hosts as free-living adult worms, display a male-biased sexual size dimorphism (e.g. Poulin, 1996e). In adult nematomorphs, competition among males for access to females in the external environment may be a key factor shaping the body size of the two sexes The situation is different in mermithid nematodes, which have independently evolved a life cycle strikingly similar to that of nematomorphs. In mermithids, females are usually much larger than males (Poinar, 1983), mirroring the pattern in other nematode taxa (Morand and Hugot, 1997). As is clear from this and preceding examples, parasite body size and sexual size dimorphism evolved independently and often differently in different parasite lineages.

4.3 EGG PRODUCTION IN PARASITES

The evolution of parasite body size is not the only aspect of parasite biology that has been plagued by misconceptions. Parasite fecundity is also commonly assumed to evolve in one direction only, toward higher egg output. Because of the massive losses suffered by infective stages during transmission, high fecundity is seen as a form of compensation ensuring that at least a few offspring will make it (Price, 1974). But surely selection should act to maximize fecundity in all organisms and not just in parasites, whatever mortality they incur at any stage of life. Contrasts between resource availability in free-living and parasitic organisms provide a better explanation for the high fecundity of some parasites (Jennings and Calow, 1975; Calow, 1983). Because parasites do not experience shortages of food, they have the resources necessary to maintain a high rate of egg production. Parasites are also often viewed as perfect examples of *r*-selected organisms, possessing many characteristics of the *r* end of the *r-K* continuum of life history strategies (Pianka, 1970; Stearns, 1992). For instance, they are generally seen as short-lived, early-maturing, small-bodied and highly prolific egg producers. Other strategies, such as the production of fewer offspring of better quality or the delay of maturity, are not considered viable options for parasites.

The following section demonstrates that the evolution of reproductive strategies in parasites has shown much more flexibility than is generally believed, and has produced a wide range of outcomes. Some correlates of egg production, and then the trade-offs among various components of reproduction, are examined. The aim throughout will be to show how strategies other than the massive production of small eggs can sometimes be adaptive.

4.3.1 CORRELATES OF FECUNDITY

Animals typically use one of two strategies of egg production. Either they produce eggs more or less continuously, or they partition their reproductive effort into discrete clutches. Estimating lifetime fecundity either from a daily rate of egg production or from an average clutch size can be difficult. Rates of egg production can vary with age, and lifetime fecundity can be quantified properly only by repeated measurements on the same individuals during their whole reproductive life. Similarly, average clutch size does not necessarily reflect lifetime fecundity, as clutch size and the number of clutches in a lifetime are not correlated (Godfray, 1987). The following discussion is based mostly on studies either of rates of egg production or of clutch sizes; the conclusions are therefore limited to these components of fecundity and not to lifetime fecundity itself.

Are transitions to parasitism followed by increases in fecundity? This question can only be answered by comparing parasitic taxa with their free-living sister taxa. Copepods and isopods parasitic on fish are no doubt more fecund than their closest free-living relatives (Poulin, 1995b, c), but a lack of data precludes any robust analysis. Again, nematodes would be an ideal group to use in settling this issue.

Once a transition to parasitism is completed, many variables will shape and constrain fecundity. In most animals, fecundity correlates with body size (Peters, 1983). Parasites are no different, with egg production varying as a function of body size both within species and across taxa. In nematodes, for instance, larger species have a higher rate of egg production (Skorping *et al.*, 1991; Morand, 1996b). Long prepatency periods, i.e. delayed maturity, result in longer adult life span and large body size in parasitic nematodes, and may have been favoured as a means of increasing rates of egg production and lifetime fecundity. Larger species of parasitic copepods, especially among those infecting fish, also produce larger clutches of eggs (Poulin, 1995b). Digeneans may be exceptions to this trend (Loker, 1983; Poulin, 1997b). For example, egg production in mammalian schistosomes does not appear to be dependent on schistosome body size (Figure 4.6). In general, though, selection for higher fecundity may drive the evolution of body size in many parasites. It may be hypothesized that high fecundity would be under stronger selection in some parasites than in others. For instance, parasites with complex life cycles may be under greater pressure to evolve high fecundity than their relatives with simple life cycles, because each egg has such an infinitesimal probability of completing the cycle. Morand (1996a), however, compared rates of egg production of nematodes with simple and complex life cycles, and found no differences.

Pressures from the host or the environment may also push or constrain the evolution of egg production. In the phylogenetic history of parasitic copepods, switching from invertebrates to fish hosts tended to be associated with increases in clutch size (Poulin, 1995b). At the same time, latitude affected egg production in copepods, creating a latitudinal gradient in copepod clutch sizes independent of body size (Poulin, 1995b). As is the case for most of the trends reported in this chapter, further comparative analyses on other taxa are needed before any generalizations can be made.

4.3.2 TRADE-OFFS AND STRATEGIES OF EGG PRODUCTION

Larger parasites tend to produce more eggs, but they can also produce larger eggs. A positive relationship exists between body size and egg size in taxa ranging from parasitic ascothoracidan crustaceans (Poulin and Hamilton, 1997) to digeneans (Poulin, 1997b). In nematodes, egg sizes show considerable variability both across species (Skorping *et al.*, 1991) and within species (Yoshikawa *et al.*, 1989), but egg size does not appear to relate to any other life-history trait. In most other taxa, though, as both egg numbers and egg size are unlikely to be maximized simultaneously, selection may have favoured different ways in which to partition reproductive effort. Indeed, trade-offs between egg numbers and egg size are observed among taxa of schistosomes (Figure 4.6) and parasitic copepods (Figure 4.7).

The strategy favoured by selection in a given taxon can range from the production of huge quantities of tiny eggs to the production of few large eggs. Parasites will not all evolve toward the many-small-eggs end of the spectrum. Again, copepods provide a good illustration of what sort of factors can shape life-history traits. Contrasts between lineages parasitic on invertebrates and their sister lineages parasitic on fish indicate that fish parasites favour the production of many small eggs, whereas parasites of invertebrates opt for few large eggs (Poulin, 1995b). Some copepods parasitic on invertebrates produce only one or two eggs per clutch, but these eggs are huge relative to the copepods' body size. Because of the greater

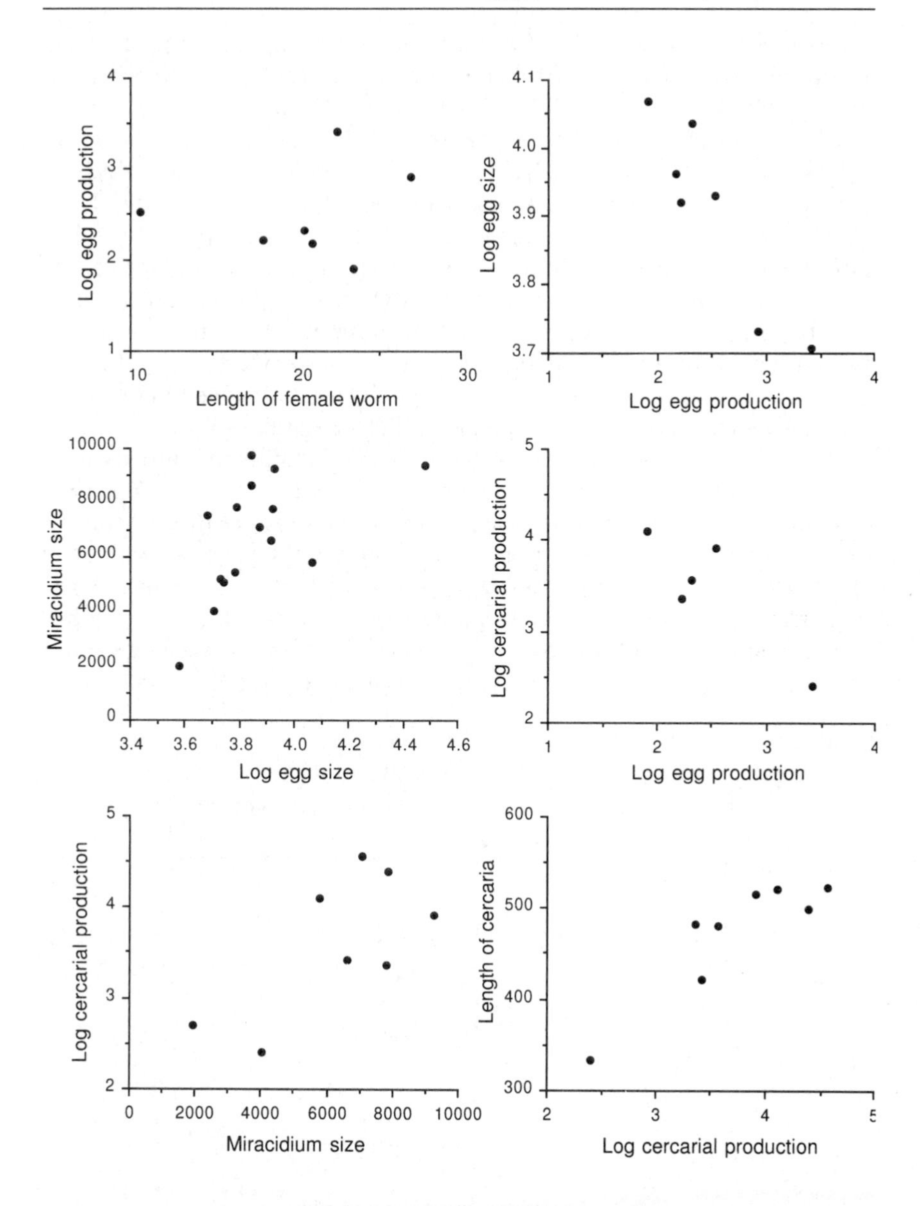

Figure 4.6 Relationships among various life-history traits of schistosome species parasitic in mammals; each point represents a different parasite species. Length of female worms is in mm, egg production is the number of eggs produced per female per day, egg size is the product of egg length (μm) and egg width (μm), miracidium size is the product of miracidium length (μm) and miracidium width (μm), cercarial production is the total number of cercariae produced per infected snail intermediate host, and cercarial length includes the tail (μm). (Data from Loker, 1983.)

difficulties associated with locating and infecting fish, selection has favoured investments in offspring numbers rather than offspring size among copepods exploiting fish hosts (Poulin, 1995b).

The probability of finding a host may also influence selection for offspring size in unionid mussels. The larvae, or glochidia, of these freshwater mussels are ectoparasitic on fish whereas adults are free-living. Bauer (1994) reported a positive relationship across mussel species between glochidial size and the number of fish species that a mussel species can use as host. Bauer did not correct for phylogenetic effects, but the relationship he reported holds, even when independent contrasts derived from a recent molecular phylogeny of unionids (Lydeard *et al.*, 1996) are used in the analysis (R. Poulin, unpublished data, 1997). Bauer (1994) suggested that the evolution of large glochidial size preceded the expansion of the host range to many fish species. Large glochidia are already at an advanced stage of development when they first attach to a fish. The parasitic life of large glochidia is a short one and they detach from the fish before a specific immune response is activated, unlike species with small glochidia in which more time must be spent on the fish. Species with large glochidia may thus have faced little pressure to specialize on a particular host taxon, and consequently they could have colonized a number of new host species. There is another explanation for the relationship,

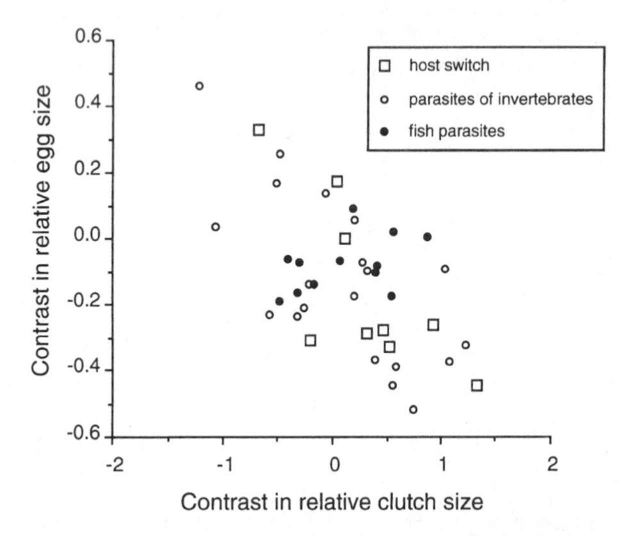

Figure 4.7 Relationship between egg size and clutch size in parasitic copepods. Each point represents a contrast between two sister branches in the copepod phylogeny. Both variables were corrected for parasite body size, which covaries with the number and size of eggs produced. The negative relationship suggests that there is a trade-off between the two components of egg production. Host switches represent cases in which one taxon is parasitic on invertebrates and its sister taxon is parasitic on fish. (Modified from Poulin, 1995b.)

however, one that involves the probability of infecting a host. In this scenario, low host specificity evolved before large glochidial size – in species that can utilize any one of a large number of host species, the probability of transmission should be higher, and investments in offspring size may have been favoured at the expense of fecundity.

Factors other than the probability of transmission can influence the trade-off between egg size and fecundity. Calow (1983) suggested that homeothermic hosts provide better growing conditions for endoparasites than poikilothermic hosts. He then hypothesized that parasites using homeothermic intermediate hosts should be selected to produce small eggs, as their larvae can grow at higher rates in the intermediate host and make up for their small initial size. Comparisons of egg size between and within cestode families support this prediction (Figure 4.8). Cyclophyllidean cestode species developing in homeothermic hosts produce smaller eggs than cyclophyllidean and pseudophyllidean species developing in poikilotherms (Calow, 1983). In digeneans, though, egg sizes appear independent of the nature of the host exploited (Poulin, 1997b), so the growth potential of the host may be of limited importance in the egg size versus egg numbers trade-off.

Furthermore, the trade-off may be influenced by environmental variables. Rohde (1993) suggested that there may be a latitudinal gradient in reproductive strategies among monogenean species. For instance, the viviparous gyrodactylids, which produce few but large offspring, are rare in warm seas but extremely diversified in northern waters. Among free-living invertebrate taxa, there is often a trend for species at low latitudes to produce more and smaller offspring than their relatives at higher latitudes (Sibly and Calow, 1986). Whether the apparent relationship in monogeneans is related to latitude or the product of some historical or phylogenetic accident remains to be determined rigorously.

Parasites with a complex life cycle in which there are two episodes of reproduction, one sexual and one asexual, may face another trade-off – between egg production by adults in the definitive host, and asexual multiplication of larval stages in the intermediate host. Mammalian schistosome species provide a good example of this possibility (Loker, 1983). Species producing eggs at a lower rate produce larger eggs; larger eggs produce larger miracidia; and larger miracidia generate greater numbers of cercariae in the snail intermediate host (see Figure 4.6). The net result is a trade-off, or a negative relationship, between egg production and cercarial production. Schistosome species either produce many eggs as adults, or many cercariae as larvae, but not both. Energy invested in egg size at the adult stage limits fecundity, but produces larger offspring better able to reproduce asexually. Surprisingly, there is a positive relationship between the number of cercariae produced and their size (Figure 4.6). This correlation may be an artefact of variation in the size of the

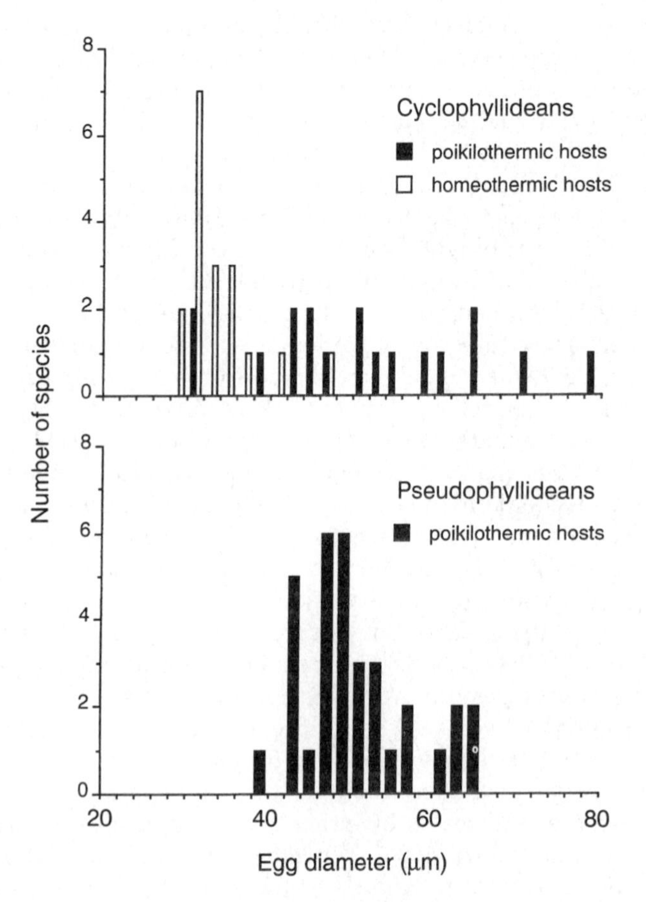

Figure 4.8 Frequency distribution of egg sizes in two orders of cestodes. Cyclophyllidean species developing in homeothermic hosts tend to have smaller eggs (average diameter, 32.3 μm) than either cyclophyllideans or pseudophyllideans using poikilothermic hosts (average diameter, 51.1 and 49.7 μm, respectively). (Modified from Calow, 1983.)

intermediate host, as cercarial production is proportional to snail size (Loker, 1983; Rondelaud and Barthe, 1987).

Asexual reproduction by larval stages has also evolved in taeniid cestodes. Theoretically, it should be favoured in highly predictable and stable environments where the genetic heterogeneity associated with sexual reproduction is not essential. The occurrence of asexual proliferation by larval stages in taeniid species is not linked with any potential indicators of predictability, such as the nature of the host or host specificity (Moore, 1981; Moore and Brooks, 1987). However, as in schistosomes, there may be a trade-off between sexual and asexual reproduction. Moore (1981) observed that taeniid species with an asexual phase are characterized by small, short-lived adults, and those with no asexual phase by large, long-lived adults. Thus adult fecundity may be maximized at the expense of larval multiplication, and *vice versa*.

4.4 SUMMARY

Life-history strategies are combinations of biological traits that have been favoured by selection because they result in higher levels of fitness than alternative combinations. In certain conditions, more than one strategy can lead to high fitness levels, and evolution can proceed in different but equally adaptive directions. Contrary to what has often been said about parasites, their evolution has followed the same diversity of paths as that of free-living organisms. Far from evolving toward small body sizes, parasites often become larger than their free-living relatives and, presumably, their free-living ancestors. Similarly, instead of invariably progressing toward higher fecundity, several parasite lineages have opted for different ways of partitioning their reproductive investments, often favouring fewer but larger offspring.

Detailed comparative studies of other life-history traits are yet to be performed. Only in nematodes have entire life-history strategies been examined from an evolutionary viewpoint (Skorping *et al.*, 1991; Morand, 1996b). Often data are unavailable, or are derived from experimental infections of laboratory hosts and not representative of natural situations. As these hurdles are overcome and as more investigations of parasite life histories are published, we will no doubt confirm that parasite strategies are as varied as those of free-living animals, and not the inevitable outcome of rigid evolutionary rules applying only to parasites.

Strategies of host exploitation

5

Most if not all definitions of parasitism involve the concept of harm, and restrict the use of the term parasites to symbionts that have a negative impact on the fitness of their hosts. In fact, the level of harm done to the host is often the main criterion used to categorize symbioses in which the host does not benefit. Symbioses fall along a continuum, from commensalism in which the host incurs no harm, to highly virulent parasitism in which host fitness is greatly reduced. This wide range of effects among contemporary symbioses has long been viewed as a series of snapshots of different associations at different stages in their evolution. Indeed, it was assumed that symbiotic associations progress through time from an initial period of adjustment, in which the symbiont causes harm to the host because it is not yet adapted to exploit it without side-effects, to a final stage of peaceful coexistence achieved when the well-adapted symbiont becomes benign to its host. Whatever the initial level of virulence, evolution was thought to proceed in one direction only, toward reduced virulence, with commensalism as the end point of parasitism.

No other evolutionary paths were seen as possible, because a parasite population evolving to become more virulent would eventually drive its host population, and therefore itself, toward extinction. This argument is clearly flawed and based on misconceptions about the process of evolution. Natural selection has no foresight; it acts blindly by favouring the genotypes among the currently available ones that lead to the highest reproductive success under the present conditions. If these genotypes also cause host mortality, and if an increase in their frequency in the parasite population could threaten the host population 50 generations down the line, they will still be favoured by selection and spread over the next generations. The realization that parasite virulence could proceed toward both lower and higher levels has changed the way we perceive the evolution of parasitism (Anderson and May, 1982; Ewald, 1983, 1994; Ebert and Herre, 1996).

The host is simply an ephemeral resource that the parasite uses to maximize its fitness, either by increasing its own probability of completing its life cycle, or by increasing its production of offspring. The way the parasite exploits the host as a resource will depend on which of the possible strategies maximizes its fitness, and not on any concern for the welfare of the host. This chapter discusses how strategies of host exploitation have evolved. The first section addresses virulence and the factors determining whether it increases or decreases over evolutionary time. The next three sections examine three well-studied specific ways in which parasites exploit hosts, using the same evolutionary framework as applied to virulence in order to determine when such strategies are adaptive.

5.1 THE EVOLUTION OF VIRULENCE

Although the notion of virulence is intuitively easy to grasp, there is no consensus on a rigorous definition of the term that specifies how virulence should be measured (Toft and Karter, 1990). In models, virulence is usually taken as the parasite-induced host mortality rate (Ebert and Herre, 1996; Frank, 1996). This is a quantitative definition of the term that allows its effect to be modelled. In practice, however, the parasite-induced mortality rate is not easy to estimate, especially if the requirement is for data from a large number of host-parasite systems for comparative analysis. In such studies, the definition of virulence ranges from subjective evaluations of the general level of harmfulness caused by a parasite (Ewald, 1995) to more quantitative measures such as parasite-induced reductions in reproductive success over the host's life time (Herre, 1993). The latter measure is probably the most significant way of estimating virulence from a host-centred evolutionary point of view.

From the parasite's perspective, however, what happens to the host as a consequence of its exploitation by the parasite may be of no importance. Selection will favour an aggressive exploitation of the host only because it leads to improved parasite fitness, not because it causes reductions in host fitness. For instance, nematomorphs and mermithid nematodes almost invariably cause the death of their insect host when they have finished using it. If the host were to recover and miraculously survive the infection, it might be tempting to view the worms as less virulent. But from the point of view of parasite evolution, selection would favour the same rate of exploitation of host resources as the subsequent death or survival of the host is of no consequence to parasite fitness. Our notion of virulence originates from medical science and is focused on impact on host fitness, whereas selection in parasites acts on rates of host exploitation irrespective of effects on host fitness.

Another reason why virulence measured as pathology can be misleading is that pathology is the result not only of parasite actions but also of host responses. Often, closely related host species incur different fitness losses from infection by the same parasite (Park, 1948; Thomas *et al.*, 1995; Jaenike, 1996b). Should we conclude that the parasite uses different exploitation strategies in different suitable host species? Of course not – some hosts are simply better able to cope with infection than others, making pathology a poor indicator of parasite strategies. In the discussion that follows, this should not be forgotten. To avoid confusion, what is meant by virulence is specified in the various models or empirical examples presented below.

5.1.1 MODELS

Parasites face a trade-off between greater exploitation and higher reproduction on the one hand, and time before the host dies on the other hand. Some parasites exploit their hosts' resources at a gentle rate, in a way that yields sustainable but not maximal benefits. Other parasites exploit their host more aggressively and cause it harm, deriving maximum benefits at the expense of long-term availability of resources. The trade-off between high reproductive rate and host survival may prevent the unchecked escalation of virulence, but will not prevent high levels of virulence from being favoured. For each parasite-host system there may be an optimal strategy of host exploitation that maximizes the parasite's lifetime fecundity.

The trade-off between the parasite's reproduction rate and host longevity can be analysed using mathematical models. In standard epidemiological models (Anderson and May, 1982, 1991), the fitness of a parasite is described by its lifetime reproductive success, R_0. Despite the limitations of this parameter for modelling the evolution of virulence (Garnick, 1992b; Ebert and Herre, 1996), it provides a good approximation of parasite fitness and is already a central feature of epidemiological theory. R_0 is obtained as follows:

$$R_0 = \frac{\beta(N)}{\mu + a + v}$$

where β is the rate at which an infected host transmits the parasite to susceptible hosts (β is dependent on host density, N), μ is the mortality rate of parasite-free hosts, a is virulence or the parasite-induced mortality rate, and v is the host recovery rate. If transmission is strictly horizontal, i.e. among host individuals of the same generation rather than from parents to offspring, strains of parasites with the highest R_0 are usually favoured. In the above equation, R_0 is highest when a equals zero. However, there is often a linkage between traits that constrains their simultaneous evolution, resulting in genetic correlations and trade-offs. A genetic linkage between β and a, for instance, can maximize R_0 at values of a different from zero (Ebert and Herre, 1996). The optimal level of virulence, i.e. the level that maximizes R_0, can be obtained using a simple rate-maximizing approach (Figure 5.1). This is the level below which improvements in the transmission rate are still possible without being annulled by the death of the host, and above which any increase in transmission rate is offset by high host mortality.

The most interesting aspect of models of virulence in the context of this chapter is their ability to make predictions about the course of the evolution of virulence in different conditions. For instance, a higher natural mortality rate of hosts due to causes other than parasites, μ in the above equation, should select for evolutionary increases in parasite virulence (Figure 5.2a). Similarly, optimal virulence should be lower in parasites that achieve relatively large increases in transmission rate, β, from small increases in virulence, a, than in parasites that obtain more modest returns from increased virulence (Figure 5.2b).

It is also possible to model various components of β, the parasite transmission rate, to see how they can affect the evolution of virulence. In directly transmitted parasites with simple life cycles, one of the main determinants of the probability of transmission (apart from host density) will be the frequency of contacts or interactions between individual hosts. In other words, the probability of transmission depends on the frequency of opportunities for transmission. If the host's social or sexual behaviour promotes contacts among individuals, transmission may be facilitated and may lead to the selection of more virulent parasites (Møller, 1996). In sexually transmitted parasites, however, host specificity is often high and the absence of alternative hosts means that there are no refuges allowing the persistence of a virulent parasite when its host population becomes

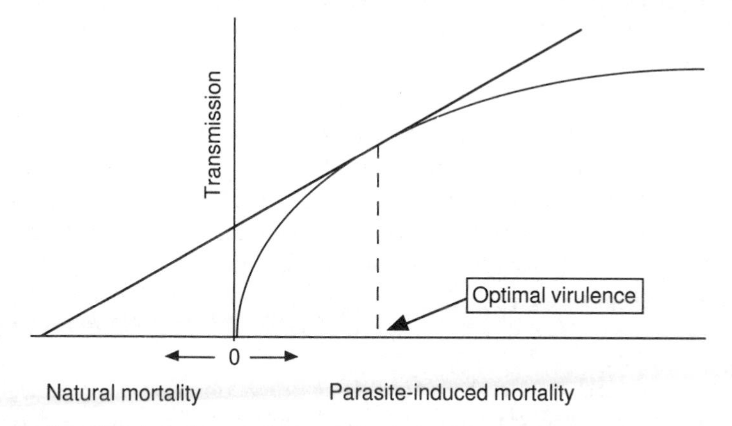

Figure 5.1 Optimal virulence of a parasite derived from the application of the marginal value theorem to the functional relationship between transmission rate, β, and the parasite-induced mortality rate, a. Parasite fitness is maximized at the point where the tangent touches the curve. (Modified from Ebert and Herre, 1996.)

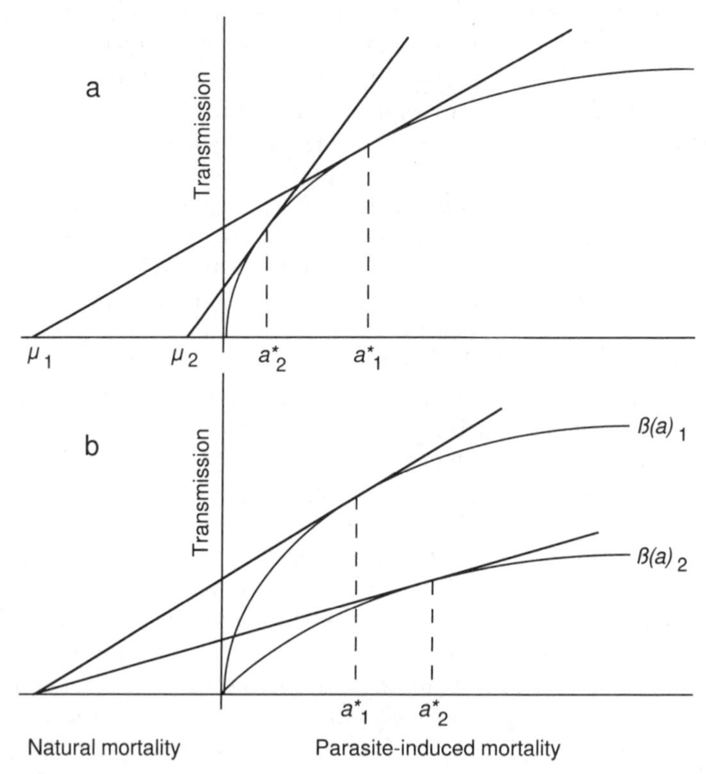

Figure 5.2 Influence of natural host mortality, μ, and the shape of the functional relationship between transmission rate, β, and parasite-induced host mortality, *a*, on the optimal level of virulence, *a**. A decrease in natural host mortality favours a decrease in parasite virulence (a), and a lower rate of returns in terms of transmission efficiency as a function of parasite-induced host mortality selects for increased virulence (b).

really small (Lockhart *et al.*, 1996). Low virulence could thus be favoured in sexually transmitted parasites even if the host is promiscuous. Strict host monogamy, the other extreme, can have similar effects: models suggest that when opportunities for transmission and contacts between the same individuals are frequent, maximum β and R_0 are achieved at low levels of virulence (Lipsitch *et al.*, 1995). Therefore, host behaviour and population structure will be important factors driving the evolution of parasite virulence.

The above are external factors that may obscure any trade-offs between maximal transmission rates and the long-term survival of the host. In fact, if the transmission route becomes independent of host health and survival, and if more propagules can be produced at high levels of virulence, evolution should proceed quickly toward high virulence

(Ewald, 1983, 1994). For instance, if transmission is achieved by vectors, selection should favour high virulence: strains producing more propagules at the expense of host health would achieve a greater transmission success than more benign strains. If the parasite is transmitted from host to host during interactions among hosts, however, selection may favour low virulence: a healthy and socially active host is essential for transmission in these parasites, and strains killing their host rapidly would become extinct. The effects of other variables acting to uncouple host health and transmission have been examined in mathematical models. Long-lived parasite propagules, for example, can under certain conditions favour an evolutionary increase in virulence, as parasites whose propagules are able to survive long periods outside the host can benefit from a rapid exploitation of the host (Bonhoeffer *et al.*, 1996). Many more models have recently been published, and have been reviewed by Anderson (1995), Ebert and Herre (1996) and Frank (1996).

The above discussion applies to horizontally transmitted parasites. In parasites that are also transmitted vertically, from parents to offspring, the optimal level of virulence should vary inversely with the proportion of the total transmission that is vertical (Mangin *et al.*, 1995; Ebert and Herre, 1996). A parasite that is always transmitted vertically and that reduces the fecundity of its host may become extinct; selection would favour more benign strains that do not jeopardize their opportunities for transmission. As transmission becomes increasingly horizontal, increases in virulence have a decreasing impact on the overall frequency of transmission opportunities and can be adaptive. In certain conditions, strictly horizontally transmitted parasites can face selection pressures identical to those acting on vertically transmitted parasites. For instance, if the prevalence of infection by a horizontally transmitted parasite is approximately 100% and is stable through time, the transmission is essentially vertical: the only uninfected and susceptible hosts available are the next generation. In this scenario, highly virulent strains of parasites would compromise the next generation of hosts and could drive their host population, and themselves, to extinction very quickly. In general, the continuum between horizontally transmitted, virulent parasites and vertically transmitted, benign parasites may hold; there are situations, however, such as when strains of different virulence compete within the host, in which different evolutionary paths are possible (Lipsitch *et al.*, 1996).

5.1.2 EMPIRICAL TESTS

Mathematical models are useful conceptual tools with which to generate testable predictions. The parameters used in models of the evolution of parasite virulence, however, have proven difficult to estimate in practice. The fitness, transmission rate and virulence of parasites (R_0, β and a in the models) are often impossible to measure in nature and can be quantified only approximately in the laboratory. Not surprisingly, then, there have been few empirical tests of some of the predictions made by the models.

Ewald (1983, 1994) has investigated the virulence of protozoans, bacteria and viruses causing diseases in humans. He defined virulence as the probability of an infection resulting in the death of the host. Comparisons among pathogens that differ in mode of transmission provide support for the predictions of theoretical models. Vector-borne pathogens, which do not need to spare the host to achieve transmission, are typically much more virulent than those transmitted directly from host to host during contact. Also, pathogens that frequently use water systems for transport are more virulent than those strictly dependent on host-to-host contacts. Medical science provides a wealth of information on human diseases and their consequences; comparative studies such as those of Ewald will be more difficult to perform with parasites of non-human animals, but are needed if we are to accept the general validity of the models' predictions.

A comparative analysis of parasites from many taxa revealed that sexually transmitted parasites are generally less virulent, i.e. less likely to induce host death, and more persistent than their relatives using other routes of transmission (Lockhart *et al.*, 1996). This is possibly a consequence of their narrow host specificity: any virulent strain of a sexually transmitted parasite causing a decline in its host population cannot take refuge in an alternative host and would become extinct along with its host.

Parasites with complex life cycles might be expected to specialize on certain hosts as resource bases, and on other hosts in the cycle as agents of transport or dispersal in space and time. Virulence could then evolve to be high in one host but lower in the next one in the cycle. Ewald (1995) gathered information on helminths with life cycles involving predation as a means of transmission, and contrasted the virulence of helminths in their definitive and intermediate hosts. He found that helminths generally have severe effects on their intermediate hosts but are usually benign to their definitive hosts. This pattern was interpreted as evidence that evolution favoured the use of intermediate hosts as resource bases and definitive hosts as vehicles for dispersal, with the debilitating effect of helminths on intermediate hosts serving to increase transmission success via predation. Ewald (1995) argues that when definitive hosts are not infected through predation, high virulence in intermediate hosts is not beneficial but instead costly; selection would instead favour the use of the definitive host as a resource base for the production of propagules. Indeed, helminths such as schistosomes that are not transmitted to the definitive host by predation tend to be virulent in their definitive host (Ewald, 1995). Although these trends are interesting, they do not present the complete picture. Virulence was measured qualitatively rather than quantitatively, and the use of intermediate hosts as resource bases by helminths transmitted *via* predation is debatable as they frequently do not grow inside the intermediate host. When they do, they can be quite virulent, as in the case of digenean larvae castrating their snail intermediate hosts (see section 5.2). Often the most detrimental effect that a parasite can have on its intermediate host is a small change in its behaviour without any apparent harm (see section 5.3). Therefore a more detailed comparative analysis is necessary to confirm Ewald's suggestion.

The influence of horizontal versus vertical transmission on the evolution of virulence has received particular attention. In an elegant laboratory study in which the probability of horizontal transmission could be controlled, selection favoured the most benign strains of a parasitic bacteriophage when transmission among bacteria hosts was restricted to vertical (Bull *et al.*, 1991). When the conditions were changed to allow the horizontal spread of the bacteriophage, the more virulent strains became predominant. Experimental evidence of this nature is difficult if not impossible to obtain for host-parasite systems with longer generation times, so we must turn to comparisons of parasites that differ in how they are transmitted. Among arthropods ectoparasitic on birds, for instance, high virulence is restricted to species that rely mostly on horizontal transmission, and low virulence is characteristic of vertically transmitted species (Clayton and Tompkins, 1994).

Parasites with both horizontal and vertical transmission stages can also shed light on the relationship between the mode of transmission and the level of virulence. Agnew and Koella (1997) found that in the microsporidian protozoan *Edhazardia aedis,* horizontally transmitted spores are associated with higher virulence in the mosquito host than vertically transmitted spores. Selection on virulence acts in opposite directions in the two transmission modes: host death is required for horizontal transmission, as spores are released into the environment only following the rupture of the host's cuticle, whereas virulence should be minimized for vertical transmission. The measures of virulence used, i.e. the degree of fluctuating asymmetry in host wing length and the size of host blood meals (Agnew and Koella, 1997; Koella and Agnew, 1997), are far from ideal but the results are interesting nonetheless.

By far the most convincing and most widely cited comparative evidence for a link between parasite virulence and opportunities of horizontal transmission comes from a study of nematodes and their fig wasp hosts in Panama (Herre, 1993, 1995). One or more gravid female fig wasps simultaneously enter a fig flower (which eventually ripens to become the fig fruit), lay their eggs and die. Their bodies can be counted to determine the number of foundress wasps per fig fruit. As the fruit ripens, the wasp offspring mature and mate inside the fig before the winged females leave to begin the cycle anew. In broods founded by a single female wasp, a count of the offspring provides a direct measure of the female's lifetime reproductive success. Each species of wasp is host to a specific species of parasitic nematode of the genus *Parasitodiplogaster*. Infective nematode larvae crawl onto newly emerged female fig wasps within figs, enter their bodies to consume them, and are carried by them to other figs. Adult nematodes emerge from the bodies of dead, infected female wasps; they mate and lay eggs in the same fig in which the host wasp has laid her eggs. The nematode eggs hatch synchronously with the next generation of wasps to repeat the cycle.

The fig wasp-nematode system allows both virulence (the reproductive success of infected female wasps relative to that of uninfected females) and opportunities for horizontal transmission (the percentage of multiple foundress broods) to be quantified for each host-parasite species pair. Herre (1993) found that the more virulent nematode species parasitize the wasp species providing the most opportunities for horizontal transmission (Figure 5.3). In wasp species where nematode transmission takes place almost always from parent to offspring, the success of the parasite is tightly linked to that of the host, and selection has favoured low levels of virulence. In wasp species offering good opportunities for horizontal transmission, there is a decoupling between host and parasite fitness, and selection favours the most virulent parasite strains (Herre, 1995). Unfortunately, whereas the lifetime fitness of hosts could be quantified, that of parasites was not. Demonstrating that highly virulent nematode species are more fecund, for instance, than less virulent species would nicely complete the story.

Ebert's (1994) study of the virulence of a microsporidian protozoan parasite of the crustacean *Daphnia magna* is the only study to demonstrate that high virulence translates into higher production of parasite propagules. Ebert (1994) infected hosts with parasites either of local origin or from distant localities. The various combinations of hosts and parasites produced different levels of parasite virulence, with high virulence associated with higher levels of spore production by the parasite (Figure 5.4). More evidence of this kind is necessary if virulence is to be considered as an adaptive parasite strategy of host exploitation.

Herre (1993, 1995) suggests that the association between nematodes and fig wasps is an ancient one, which began as a commensal, phoretic relationship in which the nematode was simply using the wasp for transport. The history of these associations is thus one of evolutionary increases in virulence. But how frequently are host-parasite associations evolving toward greater parasite virulence? This is a difficult question to answer, as it depends on the initial conditions in these various associations. Recent evidence, however, tends to refute the conventional wisdom that parasites evolve to become less virulent, and indicate that parasites in novel hosts are often less virulent or less infectious than in their original host (Ballabeni and Ward, 1993; Ebert, 1994; Dufva, 1996; Ebert and Herre, 1996).

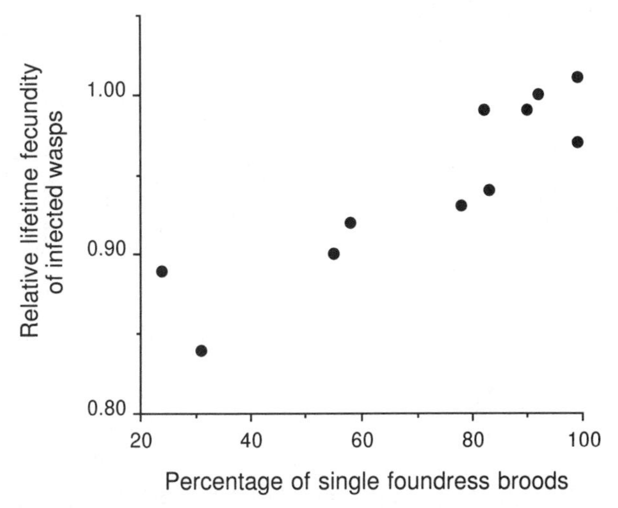

Figure 5.3 Relationship between nematode-induced reductions in lifetime fecundity and the percentage of single foundress broods among 11 Panamanian species of fig wasps. Virulence of the nematode species infecting each species of wasps can be estimated as the lifetime reproductive success of infected female wasps relative to uninfected conspecifics. A higher incidence of broods founded by multiple female wasps provides increased opportunities for nematode transmission: the more virulent nematode species are associated with wasp species presenting greater opportunities for horizontal transmission. (Modified from Herre, 1993.)

In contrast, many new host-parasite associations are characterized by high parasite virulence. One explanation may be that initially hosts have no defences against a new parasite, but may evolve some over time. For instance, the striped bass, *Morone americana*, introduced 100 years ago from the east coast of North America to the west coast was at first suffering severe pathology from infection by larvae of the cestode *Lacistorhynchus dollfusi* (Sakanari and Moser, 1990). Today, fish from west-coast populations are less debilitated and more resistant to cestode infections than fish from east-coast populations. As the bass is only one of many host species used by the cestode in the west, this is unlikely to reflect a decrease in parasite virulence: it is an evolved host response. This example illustrates how difficult it is to use host pathology as a measure of virulence if virulence is seen as a parasite strategy.

5.2 PARASITIC CASTRATION AND HOST GIGANTISM

Models of parasite virulence usually assume that achieving a high transmission rate by a ruthless exploitation of the host can only be realized at the expense of host survival. Some parasites, however, have become specialized to exploit the reproductive organs of the host. This can be

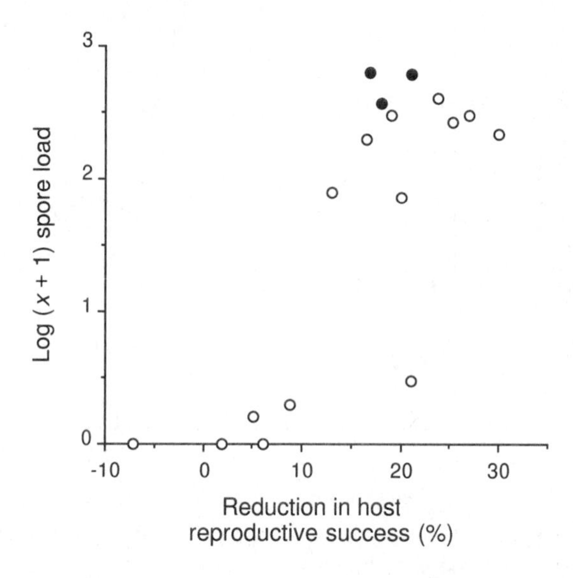

Figure 5.4 Relationship between the virulence (measured as a reduction in host reproductive success) and the spore production of the microsporidian parasite *Pleistophora intestinalis* infecting the planktonic crustacean *Daphnia magna*. Across the 17 combinations of hosts and parasites of either sympatric (filled circles) or allopatric (open circles) origin, higher virulence leads to greater production of parasite spores. (Modified from Ebert, 1994.)

achieved directly, by feeding on the gonads of the host, or indirectly either by diverting energy away from gonad development or by the secretion of 'castrating' hormones (Baudoin, 1975; Coustau *et al.*, 1991; Schallig *et al.*, 1991). Many parasitic taxa include castrators that channel energy away from host reproduction toward their own growth; the better known are digeneans in their mollusc host, and crustaceans such as rhizocephalans and epicaridean isopods in their crustacean hosts (Kuris, 1974; Baudoin, 1975). Castration does not have to be complete. In fact, 'castrated' hosts are often capable of some gamete production, and the term castration is used loosely. Whether complete or not, host castration may be the ideal strategy of host exploitation: by attacking non-vital organs, castrators do not reduce the host life span, and they can obtain a high transmission rate without trading off longevity. Obrebski (1975) modelled the benefits of castration, and found that castration is particularly advantageous to the parasite if the natural death rate of the host (μ in the models of section 5.1.1) is high. He predicted that castration would be more likely to evolve in parasite taxa exploiting hosts with short life spans, a prediction that remains untested. Jaenike (1996b) also modelled the evolution of parasitic castration and concluded that it should be favoured under a wide range of ecological conditions.

Advantages other than a normal host life span can be associated with the exploitation of host gonads. Rhizocephalans usurp the space normally occupied by the host gonads and eggs, and obtain the protection and ventilation usually provided by the host to its own offspring (Baudoin, 1975; Høeg, 1995). Also, castrated hosts often divert energy toward somatic growth instead of reproduction, and often grow to larger sizes than their unparasitized conspecifics. The increased growth of castrated hosts is not caused by the secretion of growth factors by the parasite, as in the well-known and non-castrating cestode *Spirometra mansonoides* (Phares, 1996), but rather is an indirect consequence of castration. This phenomenon is common in molluscs infected with digeneans, and can benefit the parasites as not only is their reproductive output typically proportional to host size, but also host survival is likely to be improved by investments in growth. Digeneans are also capable of targeting specific organs within the snail host; for instance they can stimulate the growth of the host's digestive gland in which they live (Théron *et al.*, 1992). The host gigantism often linked with castration is therefore seen as a parasite adaptation, whereby the host phenotype is controlled by parasite genes (Baudoin, 1975; Dawkins, 1982).

As is often the case when the label adaptation is applied to phenomena as complex as gigantism, other explanations are possible and just as valid. Gigantism could be a side-effect of the destruction of the gonads, or it could be an adaptive response of infected hosts. Minchella (1985) argues that gigantism of molluscs harbouring larval digeneans could be a host strategy to prolong survival in the hope of outlasting the infection. In molluscs not completely castrated, normal host reproduction could resume once the parasite has died. Gigantism is not the rule, and decreases in growth rates are also observed in molluscs parasitized by larval digeneans (Minchella, 1985; Mouritsen and Jensen, 1994). It may be easier to reconcile the occurrence of gigantism with life-history patterns of the mollusc hosts if gigantism is viewed as a host adaptation rather than a parasite adaptation. For instance, Sousa (1983) suggested that gigantism would be favoured as a host adaptation only in short-lived semelparous mollusc species. To date no attempt has been made to match gigantism with life-history traits among mollusc species infected with digeneans.

A recent experimental study attempted to distinguish between gigantism as a parasite strategy of host exploitation, and gigantism as a host adaptation to compensate for the effects of parasitism (Ballabeni, 1995). In the laboratory, the digenean *Diplostomum phoxini* only induced gigantism in snails from a population naturally infected with this parasite, but not in snails from another, uninfected population (Figure 5.5). This result strongly suggests that gigantism is a host adaptation that has appeared only in the host population exposed to parasites over evolutionary time.

Therefore gigantism following infection by a parasitic castrator may be an adaptive response of infected hosts to the adaptive parasite strategy of castration. However, the main benefit for parasites associated with host castration, i.e. prolonged host survival, is also not always apparent. Recent studies report contrasting results regarding mollusc survival following infection by digeneans (see Huxham *et al.*, 1993; Jokela *et al.*, 1993; Mouritsen and Jensen, 1994; Ballabeni, 1995), and the adaptiveness of castration may not be as general as is often believed.

5.3 MANIPULATION OF HOST BEHAVIOUR BY PARASITES

Host exploitation by parasites can take more subtle forms than those mentioned thus far in this chapter. Many parasites cause no measurable alterations in host growth, reproduction or survival. However, they still exploit the host in order to maximize their own transmission in a way that increases the probability that host fitness will be reduced. In parasites transmitted from the intermediate host to the definitive host by predation, parasite-induced increases in the susceptibility of the intermediate host to predation by the next host can lead to all-or-nothing fitness costs for the intermediate host. As a result of the parasite's actions, the host has a probability q of being captured by a predator and a probability 1-q of suffering no fitness cost at all. Seen from the host perspective,

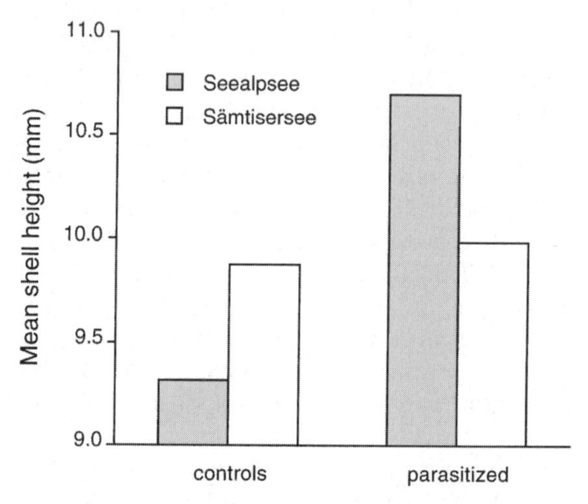

Figure 5.5 Effect of the castrating digenean parasite *Diplostomum phoxini* on the growth of its snail intermediate host, *Lymnaea peregra*. Several weeks after a laboratory infection, parasite-induced gigantism was observed only in snails from the Seealpsee Lake, where the parasite occurs naturally, and not in snails from the Sämtisersee Lake from which the parasite is absent. (Modified from Ballabeni, 1995.)

such parasites may appear non-virulent. However, they have been selected to do all they can to be ingested by their next host. Their success is absolute or nil, never partial. Because of the probabilistic nature of their transmission, their virulence, as traditionally measured by effects on host fitness, will often appear minimal. This is yet another situation in which a parasite-centred measure of host exploitation would be preferable to actual virulence in order to investigate parasite evolution.

5.3.1 ADAPTIVE MANIPULATION?

Parasite-induced alterations in host behaviour or coloration have been reported in a wide range of protozoan and metazoan parasites, most of which have complex life cycles (Holmes and Bethel, 1972; Moore and Gotelli, 1990; Moore, 1995; Poulin, 1995e). The simplest explanation for these changes is that they are non-adaptive, coincidental side-effects of parasitic infections. The very complex nature of some parasite-induced behavioural changes, however, often suggests that they could benefit either the host or the parasite, and therefore be adaptations. Examples include apparent increases in the vulnerability of intermediate hosts to predatory definitive hosts, but also alterations of host behaviour in systems where transmission takes a different route. Vector-borne parasites can render their vertebrate hosts more susceptible to vector feeding (Day and Edman, 1983), and can affect the feeding behaviour of their vector hosts (Moore, 1993). Fungal parasites of insects can make their hosts climb trees and die perched in an optimal position for the dispersal of fungal spores by wind (Maitland, 1994). Some digeneans make their snail intermediate hosts change microhabitat and move to ideal sites for the release of cercariae (Curtis, 1987; Lowenberger and Rau, 1994). Mermithid nematodes make their insect hosts enter water, where the adult worms must emerge (Vance, 1996). Usurpation of host behaviour leading to increased transmission of parasites is thus common in many parasite taxa, and is also observed in insect parasitoids (Fritz, 1982; Brodeur and McNeil, 1989; Brodeur and Vet, 1994).

Alterations in host behaviour following parasitic infection are often exactly what we would expect to see if the host were to start acting in a way that benefits the parasite. They appear to be adaptations rather than mere pathological side-effects. In fact, changes in host behaviour after infection may be perfect examples of the extended phenotypic effects of genes, which reach out of their own bodies to influence other organisms (Dawkins, 1982, 1990). The fact remains, however, that the adaptiveness of most parasite-induced changes in host behaviour is yet to be demonstrated convincingly (Moore and Gotelli, 1990; Poulin, 1995e). The most important criterion that must be met by any behavioural change to be labelled a parasite adaptation is that it must lead to improved parasite transmission or fitness. Definitive hosts may have little to gain by actively avoiding infected prey; infected prey with altered behaviour are easier to capture, and typically the adult parasite causes little pathology in the definitive host (Lafferty, 1992). Despite these arguments, empirical evidence of fitness benefits is necessary. Only a handful of studies have shown that infected intermediate hosts are more susceptible than uninfected individuals to predation by their definitive hosts under field conditions (Figure 5.6). Results of laboratory studies usually (but not always) show similar trends (Poulin, 1995e). Nevertheless, only a fraction of known parasite-induced changes in host behaviour have been tested for fitness effects.

Other lines of evidence can suggest that behavioural changes in hosts are parasite adaptations in the absence of demonstrated fitness benefits. Complex traits that suggest a purposive design are less likely to evolve by chance and are therefore more likely to be the adaptive products of selection (Poulin, 1995e). Consider the well-known digenean *Dicrocoelium dendriticum* which must be transmitted from an ant to a grazing sheep by ingestion (Carney, 1969; Wickler, 1976). *Dicrocoelium* causes infected ants to climb to the tip of grass blades and stay there waiting for sheep. Another digenean, *Leucochloridium* spp., alters the shape, size and col-

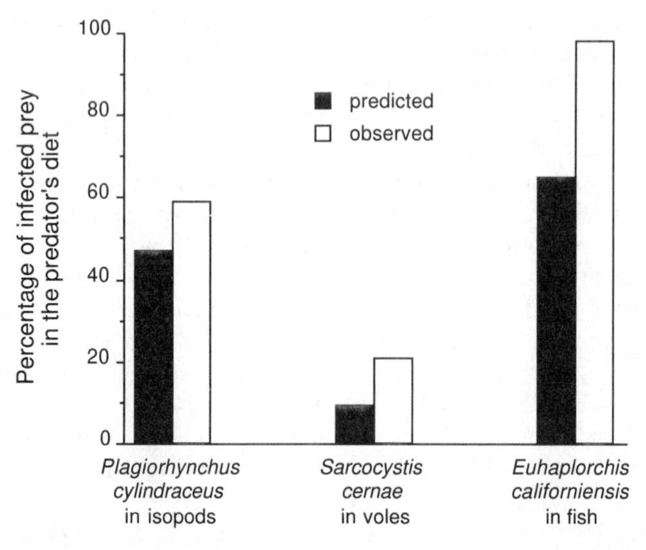

Figure 5.6 Predicted and observed percentage of parasitized prey in the diet of definitive hosts of three parasites transmitted by predation. Predicted values are the prevalences of parasites in the populations of available prey; observed values are from actual prey captures under natural conditions. The acanthocephalan *Plagiorhynchus cylindraceus* passes from its terrestrial isopod intermediate host to starlings; the protozoan *Sarcocystis cernae* goes from voles to kestrels; and the digenean *Euhaplorchis californiensis* is transmitted from fish to piscivorous birds. (Data from Moore, 1983; Hoogenboom and Dijkstra, 1987; Lafferty and Morris, 1996.)

oration of the tentacles of its snail intermediate host, and causes them to pulsate in response to light (Wesenburg-Lund, 1931). This seems an excellent way to attract the attention of bird definitive hosts. Fitness benefits for the parasites have not been demonstrated in either of these two digeneans, but given the close fit between the presumed function of these behavioural changes and the parasites' life cycles, it is reasonable to assume that such complex alterations of host behaviour and coloration are extremely unlikely to arise by chance. Note, however, that the argument can also be misused: a strong correspondence between the phenotypic expression of a host-behaviour modification and the design that an engineer might specify for the function of transmission is seen by some as evidence of divine creation (Smith, 1984).

The majority of known parasite-induced behavioural changes are simple increases or decreases in the proportion of time during which infected hosts perform certain behaviours with respect to uninfected hosts (Poulin, 1995e). Such simple alterations are more likely to be accidents of chance or pathological side-effects than the complex changes in host behaviour induced by *Dicrocoelium* or *Leucochloridium*, and caution is needed when evaluating their adaptiveness. For instance, small changes in host activity caused by parasites can go in opposite directions and have different effects on transmission among closely related host-parasite systems (Figure 5.7). Increases in activity are just as likely to be observed as decreases in activity, and either change may affect transmission success. Variability in parasite-induced changes in host behaviour also exists among strains of hosts (Yan *et al.*, 1994), softening any conclusion about the adaptiveness of these changes.

One indirect way of demonstrating that parasite-induced behavioural changes are parasite adaptations would be to identify the mechanism causing the changes. The presence of a specialized structure or organ in the parasite responsible for the manipulation of host behaviour would not be the product of random evolution. At the very least, if their effect on parasite fitness cannot be verified directly, simple changes in host behaviour should coincide with the onset of parasite infectivity to the next host (Bethel and Holmes, 1974; Hurd and Fogo, 1991; Poulin *et al.*, 1992). In addition, they must be shown not to be simple host responses (Hechtel *et al.*, 1993) or pathological reactions (Robb and Reid, 1996) before being considered potential adaptations.

Adaptations can also be recognized at the macroevolutionary scale. If different parasite lineages with similar life cycles have independently evolved the ability to cause identical alterations in host behaviour, then surely this ability is adaptive. Comparative analyses of parasite-induced behavioural changes can be powerful tools for understanding the evolution of host exploitation strategies. To date, only one study has examined host behavioural changes in a phylogenetic context. Moore and Gotelli

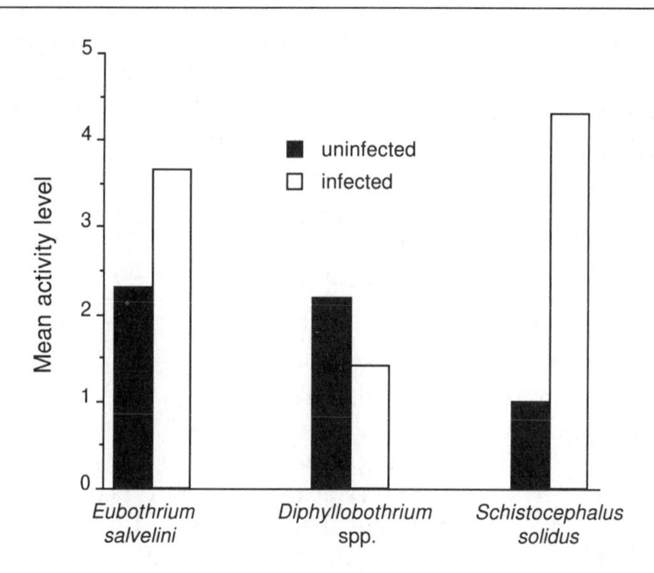

Figure 5.7 Effect of infection by three cestode species on the activity of freshwater planktonic copepods (*Cyclops* spp.). The activity of copepods harbouring *Eubothrium salvelini* was measured as distance travelled (× 10 mm); that of copepods harbouring *Diphyllobothrium* spp. as hops per second; and that of copepods infected with *Schistocephalus solidus* as the number of visits to different depth zones. Infections with *E. salvelini* and *Diphyllobothrium* spp. resulted in greater susceptibility of the copepod host to predation; infections with *Schistocephalus solidus* had no effect on risk of predation. (Data from Poulin *et al.*, 1992; Pasternak *et al.*, 1995; Urdal *et al.*, 1995.)

(1996) showed that the ability of the acanthocephalan *Moniliformis moniliformis* to alter the behaviour of several cockroach species showed no concordance with a cockroach phylogeny derived from morphological data. In other words, behavioural alterations caused by the same parasite are not legacies from an ancestor but are derived traits that have arisen repeatedly and independently in different host lineages. This suggests that alterations of host behaviour can be adaptive, but the truly interesting test would be to compare the ability of different parasite species to manipulate host behaviour. Convergence of traits in different organisms under similar selective pressures can be instructive, but so can differences in these traits if they relate to differences in selective pressures. Within parasite taxa, the ability of different species to alter host behaviour varies greatly when measured as the magnitude of the induced changes (Poulin, 1994b). In some taxa, host manipulation seems widespread and well-developed (e.g. in acanthocephalans; Moore, 1984), whereas in other taxa it is more variable (see Figure 5.7). These latter taxa offer ideal opportunities for further comparative analyses of parasite-induced alterations in host behaviour.

5.3.2 EVOLUTION OF HOST MANIPULATION

From a theoretical perspective, variability in the ability to manipulate host behaviour is expected among parasite lineages. Manipulation may be costly, and to be favoured by selection, the fitness benefits must outweigh the costs incurred. Some parasites may produce changes in host behaviour without incurring any costs, simply by being in the right organ by chance. Most alterations of host behaviour, however, appear to result from interference with host neurochemistry that may involve the secretion and release of hormones and other neurochemicals by the parasite (Helluy and Holmes, 1990; Holmes and Zohar, 1990; Hurd, 1990; Thompson and Kavaliers, 1994; Maynard *et al.*, 1996). The development of specialized glands or tissues for the production of these chemicals must be costly. In some systems, changes in host behaviour that benefit the parasite may be fortuitous and cost-free; in general, however, costs of manipulation exist even if they are difficult to quantify (Poulin, 1994a). Any energy invested in host manipulation will not be available for growth, reproduction, or fighting the host's immune system. These trade-offs suggest that selection will not always favour large investments in host manipulation. Investments in manipulation, or manipulation effort (ME), will tend toward an optimal value at which parasite fitness is maximized. Under some ecological conditions, low values of ME will be favoured, and the associated changes in host behaviour may be very small.

It is possible to make general predictions about the optimal ME, or ME*, expected in different systems (Poulin, 1994a). Even with no investment in host manipulation (ME = 0), the transmission success of the parasite is unlikely to be nil. For instance, some infected intermediate hosts would still be eaten by predatory definitive hosts even in the absence of manipulation by the parasites. Typically, prevalence of infection is extremely low among invertebrates serving as intermediate hosts for helminths (see review in Marcogliese, 1995). However, because definitive hosts ingest large numbers of prey over time, chances are that random selection will result in some infected intermediate hosts being captured. Investments in manipulation (ME > 0) will only increase the probability of transmission above the passive transmission rate, or p (Figure 5.8). The net benefits of manipulation correspond to the difference between the transmission rate achieved through manipulation and the passive transmission rate. As ME increases, the rate of increase in the probability of transmission is likely to follow a law of diminishing returns (Figure 5.8): small investments yield greater returns per investment unit than large investments. The investment favoured by selection, ME*, will depend on p, on the exact shape of the curve in Figure 5.8, and on the cost of manipulation.

These variables will in turn be influenced by the ecological conditions prevailing in the host-parasite system. Consider the mean infrapopula-

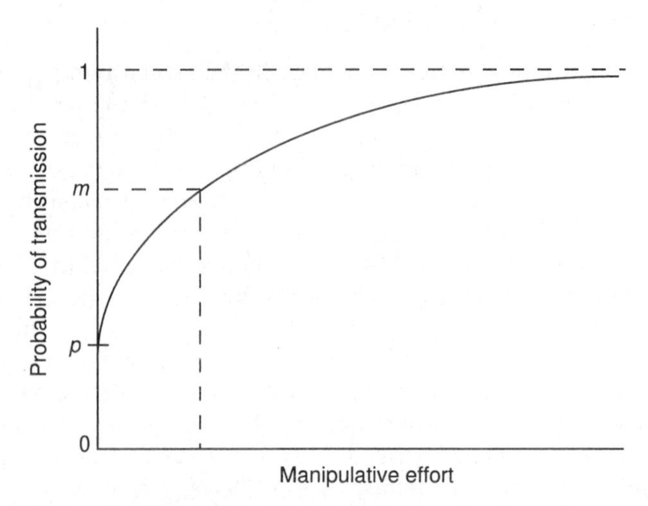

Figure 5.8 Probability of parasite transmission as a function of the manipulative effort made by the parasite, i.e. the energy invested in altering host behaviour. The probability ranges from p, the passive transmission rate, to the hypothetical maximum of 1. The benefits of manipulation equal $m - p$, or the difference between the realized transmission rate and the passive transmission rate. Increments in manipulative effort result in greater increases in the probability of transmission when the current manipulative effort is small than when it is large. (From Poulin, 1994a.)

tion size, or the mean number of conspecific parasites sharing a host individual, and its effect on ME*. In systems where infrapopulations are small, e.g. an average of a single parasite per host, the task of modifying host behaviour cannot be shared and selection may favour a high ME*. In contrast, if a parasite is likely to share its host with several conspecifics, the costs of manipulation can be shared: if the impact of manipulation by several individual parasites is cumulative, then small ME* values can achieve the same transmission rate obtained by the single parasite doing all the work on its own. Of course, because of the aggregated distribution of parasites among their hosts (see Chapter 6), the mean infrapopulation size is meaningless, as it does not represent the infrapopulation size experienced by the majority of parasites. Selection cannot, therefore, favour a fixed ME* based on infrapopulation size as it is too variable. One solution may be to favour higher ME* values than those expected strictly from a consideration of mean infrapopulation size. Parasites finding themselves crowded with several conspecifics in their host would pay a higher cost than necessary, but this strategy would guarantee a decent chance of transmission to parasites ending up alone in their host (Poulin, 1994a). Another, better strategy would be to develop a plastic ME rather than a fixed ME*. This would allow a parasite to adjust its investment in manipulation to the conditions it happens to encounter rather than follow an inflexible investment plan.

In populations where there is a more-or-less fixed ME*, what prevents some cheating parasite from paying fewer costs (ME < ME*) while still obtaining all the benefits by relying on the action of its conspecifics? Cheating genes could spread quickly through a parasite population when they first appear. Eventually, however, the need for some honest, manipulative parasites to do the work would place limits on the spread of cheating genes. The ME* strategy would return as the most profitable strategy (Poulin, 1994a). Cheating may not be a viable option among conspecifics, but taking advantage of another species' ability to manipulate hosts should be favoured by selection. Parasites often share intermediate and definitive hosts with other parasite species; if one species is an efficient manipulator of host behaviour, then the others may make no investment in manipulation (ME* = 0) and obtain a high probability of transmission simply by riding along. Thomas *et al.* (1997) suggest that this may happen in two digeneans using the same crustacean as intermediate host and the same bird species as definitive hosts. One species is a manipulator and the other is not; however, the non-manipulator species preferentially infects intermediate hosts harbouring manipulators. It does so with no obvious costs and reaps the benefits provided by the manipulator species.

In many systems, entire infrapopulations may consist of related individuals. For example, most if not all cercariae of *Dicrocoelium dendriticum* ingested by an ant are derived from the same sporocyst inside the snail serving as first intermediate host. In other words, all cercariae come from the same egg laid by a single adult parasite (Wickler, 1976). In the ant, only one larva attaches to the ant's suboesophageal ganglion and induces a behavioural change in the host. This parasite dies and is not transmitted when the ant is ingested by the sheep definitive host, a good example of the cost of manipulation. Nevertheless, this individual benefits in terms of inclusive fitness, as its sacrifice helps its identical kin complete their life cycle. In this case, selection has favoured one individual paying the full cost of manipulation rather than the sharing of the cost among relatives. Because of the asexual multiplication of many parasite larvae, infrapopulations of close relatives may be common in intermediate hosts, and various methods of paying the costs of manipulation may have been favoured over the all-or-nothing strategy of *D. dendriticum*.

Many factors other than infrapopulation size or composition can affect ME* (Poulin, 1994a). Transmission by predation from an intermediate host to a definitive host is an unlikely event in most cases. To counter the odds, parasites can use several strategies. They can produce more eggs at the adult stage, or they can increase their longevity in the intermediate host in order to wait for the unlikely predation event. If there are trade-offs between ME and either adult fecundity or longevity in the intermediate host, lower values of ME* may be expected. Different strategies, involving different combinations of manipulation, fecundity, larval longevity, etc., may all result in similar transmission rates. Which of these strategies actually evolves depends on the genetic variability of the different traits and on various other constraints. The important point is that we cannot expect manipulation of host behaviour to be the rule in systems where such manipulations appear adaptive. Just as with high virulence, host manipulation is only one of many strategies that may be selected. Perhaps the only approach available to investigate the evolution of host manipulation is to relate either its occurrence or the degree of its expression (ME*) to some of the many life-history and ecological variables to which it may be linked (Poulin, 1994a). Comparative analyses across species, once again, are the only way to general answers.

5.4 MANIPULATION OF HOST SEX RATIO BY PARASITES

Among parasites transmitted vertically, several protozoans use host gametes to move from one host generation to the next (Canning, 1982; Dunn *et al.*, 1993). This transmission mode is also common in viruses and bacteria (Werren *et al.*, 1988; Hurst, 1993). These parasites inhabit the cytoplasm of host gametes and are passed on to host offspring at reproduction. Because of the difference in size and in the amount of cytoplasm between ova and sperm, the parasites are generally transmitted only from infected female hosts to their offspring. In male hosts, if horizontal transmission does not occur, the parasites are at an evolutionary dead end. Of the potential strategies of host exploitation open to maternally transmitted parasites, biasing host investment in favour of female offspring appears to be a common option (Hurst, 1993).

Parasites capable of increasing the proportion of host offspring that are of the transmitting (female) sex and thus not dead ends are called sex-ratio distorters. They manipulate the sex ratio of host offspring in three ways (Dunn *et al.*, 1995). Firstly, they can kill the male offspring of their host, such that more resources are diverted toward female offspring that serve as routes of transmission. The release of infective stages from dead hosts for horizontal transmission can be another benefit obtained from killing male offspring. Secondly, they can distort the primary sex ratio of the host, by converting genotypic males into genotypic females. Thirdly, they can feminize the male offspring of the host, turning genotypic males into functional phenotypic females. By turning male hosts into females capable of transmitting the infection, the parasite increases its base for transmission to the next generation of hosts. Feminization can be achieved by the suppression of androgen gland development during embryogenesis or by other physiological manipulations; whatever the mechanism, feminization can produce strongly female-biased sex ratios among the offspring of infected hosts (Figure 5.9).

The effect of a feminizing parasite on host sex ratio will depend on the host's intrinsic primary sex ratio, the parasite's transmission efficiency from the mother host to her offspring, and the parasite's feminizing influence (Figure 5.10). Whereas natural selection should favour increases in the parasite's transmission and feminizing efficiency (t and f in Figure 5.10), it may at the same time favour hosts producing more offspring of the rarer sex (Hatcher and Dunn, 1995). Despite the evolution of compensatory sex ratios in the host population, increases in the transmission and feminizing efficiency of the parasite will be favoured so long as they do not affect other components of the parasite transmission success, such as host fecundity. When transmission and feminization attain their maximum value ($t = 1$ and $f = 1$ in Figure 5.10), the parasite may drive its host population, and hence its own population, extinct because of a lack of male hosts. In fact, even at values of t and f less than unity, it is theoretically possible for the parasite to drive its host population to extinction, for instance if the host

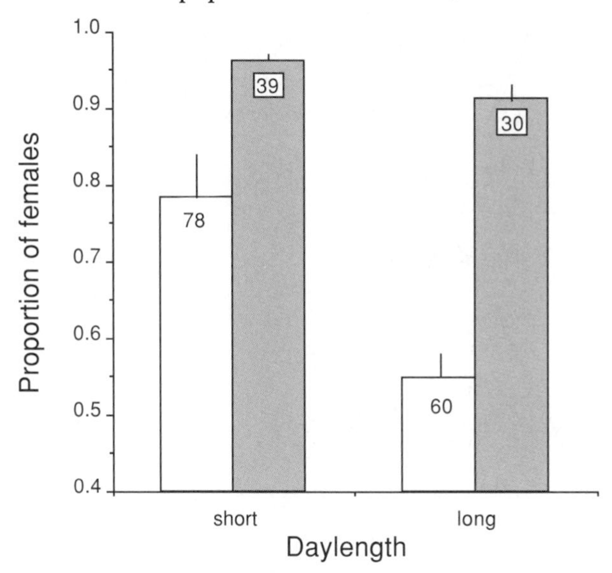

Figure 5.9 Mean proportion of female offspring in the broods of the female amphipod *Gammarus duebeni* either uninfected (open bars) or infected by the transovarially transmitted microsporidian protozoan *Octosporea effeminans* (shaded bars). Standard errors and numbers of broods in each treatment are also shown. Female amphipods were raised under different daylength conditions, because photoperiod is an environmental factor known to influence sex determination in *G. duebeni*. Whatever the photoperiod, parasitized amphipods produced much more female-biased broods than uninfected ones, with more than 90% of offspring of parasitized mothers becoming females due to the feminizing effect of the protozoan. (Modified from Dunn *et al.*, 1993.)

population is small. Thus parasitic sex-ratio distortion is a risky strategy for host exploitation and parasite transmission: traits that maximize transmission also jeopardize the host population. The relatively short list of parasites known to distort host sex ratio may reflect a lack of research effort. It may also be a consequence of the risks of adopting this strategy, the highly successful taxa having become extinct without leaving a trace.

5.5 SUMMARY

The past few years have witnessed much progress in our understanding of parasite evolution. It is now generally accepted that parasites do not

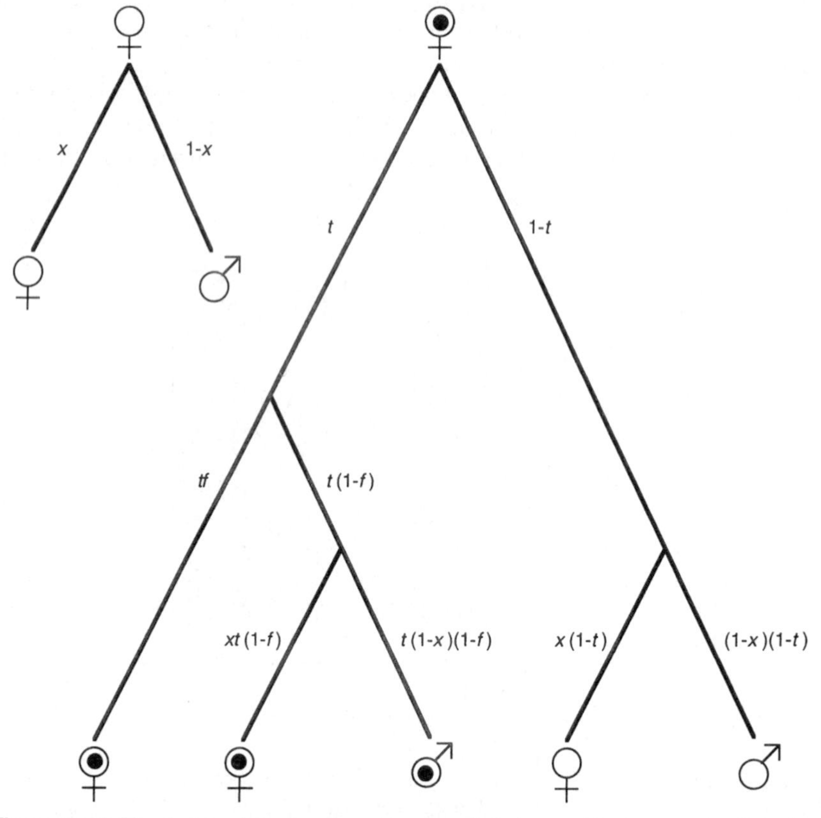

Figure 5.10 Effect of feminization by a transovarially transmitted, sex ratio-distorting parasite on the proportion of male and female offspring produced by uninfected and infected female hosts. Infected hosts are indicated by a black dot. The proportion of female offspring in the brood of uninfected female hosts equals x, which is usually approximately 0.5 in uninfected populations. Broods of infected female hosts are likely to have female-biased sex ratios, determined by t, the proportion of the brood that inherit the parasite, and f, the proportion of infected offspring that are feminized by the parasite. (Modified from Hatcher and Dunn, 1995.)

invariably evolve to become harmless to their host. A diversity of strategies for the exploitation of the host have flourished among the many taxa of parasites, such that parasites cause anything from undetectable to drastic changes in host fitness. With the data currently available and with rapid advances in the resolution of parasite phylogenies, we can start using comparative studies to elucidate the causes behind the variation in patterns of virulence, castration, and sex-ratio or behaviour manipulation.

Yet we may need to modify our approach to these problems. Currently, we investigate the evolution of parasite strategies by investigating their effects on hosts. To quantify how quickly parasites exploit hosts as food resources, we quantify the pathology incurred by the host rather than the actual exploitation rates and how they relate to parasite fitness. To examine how parasites manipulate their hosts, we measure changes in host behaviour instead of quantifying how much energy is invested in manipulation by parasites or how it translates into fitness gains. What happens to the host is more a reflection of how well the host can cope with a parasite than of what the parasite does. A final example shows how focusing on the effects of parasites on hosts can be misleading when we are really interested in parasite adaptations. Following infection with metacercariae of the digenean *Plagioporus*, coral polyps become distended, causing coral colonies to suffer reduced growth (Aeby, 1992). Butterfly fish, the parasite's definitive host, feed preferentially on parasitized polyps, which appear pink and differ in coloration from healthy polyps. This looks like a case of host manipulation by the parasite, in which the parasite benefits and the host suffers from lower growth and higher predation risk. However, predation enhances the regenerative capabilities of the coral colony, such that the net effect of parasitism on coral fitness may be positive (Aeby, 1992). Focusing exclusively on fitness effects on the host, or on the changes in the susceptibility of hosts to predation, would give us a biased picture of what strategy is really employed by the parasite. Clearly, studying parasites through their impact on hosts is not the best way to proceed.

Parasite aggregation: causes and consequences

6

The previous chapters have focused on properties of individual parasites, somewhat variable among members of the same species but shared by all. The next unit of organization is the population, consisting of conspecific parasites interacting and coexisting in time and space. Because of their complex life cycles, different members of parasite populations occupy completely distinct habitats. For instance, infective larval stages may be in the external environment, immature stages in intermediate hosts, and adults in definitive hosts. This fragmentation of the parasite population complicates its study unless the different stages are examined separately. This chapter looks at the distribution of parasites in space; it will discuss mainly parasitic stages in hosts as opposed to free-living infective stages, as parasitic stages have received the most attention from parasitologists.

Free-living animals are not evenly distributed across their geographical range; the abundance of a species varies in space in response to variations in the suitability of the habitat. The abundance of a species is usually highest in the parts of the range where conditions are near optimal, resulting in an uneven distribution of abundance across the geographical range (Hengeveld and Haeck, 1982; Hengeveld, 1992). The habitat of parasites differs from that of most free-living animals: it is not spatially continuous but consists of discrete cells or islands. Hosts represent patches of suitable habitat in an otherwise inhospitable environment. Parasites are not uniformly distributed among these patches, so that some patches contain many more parasites than the average, and others harbour fewer. Typically, parasites are aggregated among the available hosts, such that most host individuals harbour no or few parasites and few hosts harbour many parasites. Parasitologists often use the word overdispersion to describe this uneven distribution. The word is confusing to many biologists, suggesting a wide dispersion rather than clumping. Here the term aggregation is preferred, and will be used as synonymous with overdispersion: it is intuitively easier for a non-parasitologist to form a mental picture of the distribution of parasites when aggregation is mentioned.

The parasite individuals of the same species inside the same host individual form an infrapopulation. Infrapopulations are subsets of the entire parasite population, or suprapopulation. In terms of infrapopulations, parasite aggregation means that most infrapopulations will be small and only a few infrapopulations will be large (Figure 6.1a). Reformulating this with respect to parasite individuals is more complicated: the total number of parasites belonging to many small infrapopulations may be roughly equal to the total number of parasites belonging to few large infrapopulations (Figure 6.1b). From an individual point of view, there may be two equally likely types of infrapopulations, small ones and large ones, in which an individual parasite may end up. This will have ecological and evolutionary consequences. These consequences, as well as the causes of aggregation, will be discussed later, following some remarks regarding the measurement of parasite aggregation.

6.1 MEASURING PARASITE AGGREGATION

The distribution of parasites is not static over time. It is the product of processes that are not constant in time, and therefore the distribution of parasites is a dynamic phenomenon. No single measure can capture the distribution. An estimate of aggregation obtained from a sample of hosts and parasites is merely a snapshot of an intricate and ever-changing distribution. Whatever measure one chooses to quantify aggregation, the dynamic nature of the distribution should never be overlooked.

6.1.1 INDICES OF AGGREGATION

The simplest and most commonly used measure of parasite aggregation is the ratio of the variance to the mean number of parasites per host. If parasites are distributed at random among their hosts according to a Poisson distribution, the mean number of parasites per host will equal the variance. A variance-to-mean ratio greater than unity indicates a departure from randomness and a tendency toward aggregation, and aggregation levels increase as the value of the ratio increases. This index of aggregation is easy to compute and to understand, which explains its popularity. Using it one finds that parasite aggregation is very much the rule (Dobson and Merenlender, 1991; Shaw and Dobson, 1995). However, this index is not without problems. There has been much confusion regarding the inclusion of uninfected hosts in computations of the mean and variance: technically they should be included, but often they are not. A more important issue is the perspective of aggregation that this measure gives us. The ratio of the variance to the mean number of parasites

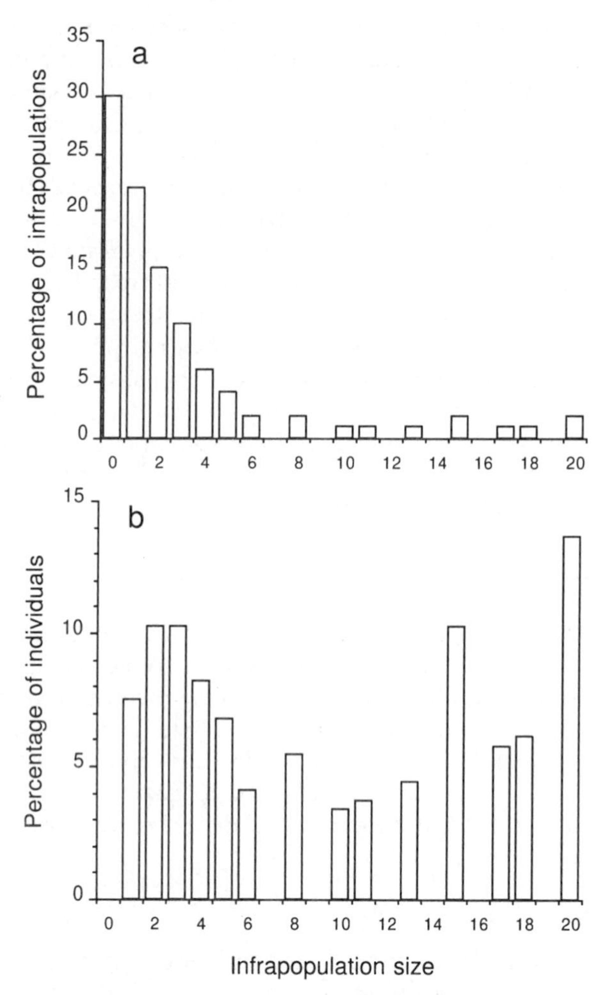

Figure 6.1 Two representations of the same hypothetical distribution of parasites among hosts. From a host-centred perspective, we can focus on the frequency distribution of numbers of parasites per host, or infrapopulation sizes (a). In addition, from the perspective of parasite individuals, we can instead look at the frequency distribution of individuals among infrapopulations of different sizes (b). Less than 25% of hosts harbour more parasites than the average number of parasites per host (3.0, including uninfected hosts). About one-third of infrapopulations are larger than the average infrapopulation size (4.2, excluding uninfected hosts). However, almost two-thirds of the parasites occur in infrapopulations larger than the average infrapopulation.

per host quantifies the variability in intensity of infection among hosts, or the variability in infrapopulation sizes including empty infrapopulations. This host-centred view may not be appropriate when we are interested in parasite individuals. Put differently, the variance-to-mean ratio characterizes the distribution in Figure 6.1a but not that in Figure 6.1b, although these are two representations of the same parasite distribution.

Ever since Crofton (1971) suggested it, fitting the negative binomial distribution to observed parasite distributions has also become common practice. Just as the Poisson distribution exemplifies a perfectly random distribution, the negative binomial is the statistical representation of aggregation. Its use is not always straightforward (see Grafen and Woolhouse, 1993), but in general when the observed parasite distribution does not depart from the frequencies predicted by the negative binomial, the parasites are aggregated. Such is the fit between observed distributions and the negative binomial that any discrepancy is often taken as evidence of parasite-induced host mortality – heavily infected hosts exist but are missing from samples because of their high mortality (Gordon and Rau, 1982; Adjei *et al.*, 1986; Rousset *et al.*, 1996). The parameter k of the negative binomial can serve as an index of aggregation: as k tends toward zero, aggregation increases, but at values of $k = 8$ or more the negative binomial converges with the Poisson series (Southwood, 1978) and the parasites are randomly distributed. Using k as a measure of aggregation, one finds aggregation is by far the predominant pattern across host-parasite systems (Anderson, 1982; Shaw and Dobson, 1995). Like the variance-to-mean ratio, however, the use of k as an index of aggregation is not without problems. The parameter is difficult to compute, and good approximations can be obtained only with the maximum likelihood method of Bliss and Fisher (1953); other mathematical shortcuts provide unreliable estimates. The parameter k is also not very sensitive to the tail of the distribution, i.e. to the heavily infected hosts, and does not change much as their numbers vary (Scott, 1987). Finally, comparisons of aggregation levels between different samples using k as an index may be totally unreliable because k is highly dependent on the mean number of parasites per host, which is likely to vary among samples (Scott, 1987). Thus k is far from an ideal index of aggregation.

Another index of aggregation is the patchiness index of Lloyd (1967). This measure has remained mostly ignored by parasitologists despite its interesting properties. It equals the mean number of conspecifics sharing a host with a parasite, calculated across all parasite individuals in a sample, divided by the mean number of parasites per host. Or, it equals the mean of the distribution in Figure 6.1b, minus one, divided by the mean of the distribution in Figure 6.1a. Lloyd's patchiness index relates to k as well, and can be computed as $1 + 1/k$. It is meant to represent the average aggregation experienced by individual parasites, and measures how much an individual is 'crowded', on average, compared to what it would experience had the parasite population been randomly distributed. In contrast to the variance-to-mean ratio and the parameter k, the patchiness index gives more weight to individual parasites and less to infrapopulations.

A more recent measure, the index of discrepancy, D, quantifies aggregation as the departure between the observed parasite distribution and the hypothetical distribution in which all hosts are used equally and all parasites are in infrapopulations of the same size (Poulin, 1993). It is computed as follows:

$$D = 1 - \frac{2\sum_{i=1}^{N}\left(\sum_{j=1}^{i} x_j\right)}{xN(N+1)}$$

where x is the number of parasites in host j (after hosts are ranked from least to most heavily infected) and N is the number of hosts in the sample. The index is represented graphically in Figure 6.2. Plotting the cumulative number of parasites as a function of the cumulative number of hosts always produces a concave curve: the more concave the curve the greater the degree of aggregation. The index of discrepancy is simply the relative departure between this curve and the straight line that would be obtained if the parasite distribution were perfectly uniform (Figure 6.2). In contrast to previous indices, the index of discrepancy has a finite range of values. The minimum value is zero, when the curve falls on the straight line and when there is no aggregation. The maximum value is one, when all parasites are in the same infrapopulation, i.e. when aggregation is at its theoretical maximum. This limited range of values facilitates comparisons between distributions that vary in prevalence or mean number of parasites per host. The index is a measure of relative discrepancy, in which each distribution is compared to its own theoretical uniform counterpart. This property makes it reliable for comparisons of aggregation levels among samples.

There are other ways of quantifying aggregation, some of which have more specific uses. For example, Boulinier *et al.* (1996) have described a method for measuring aggregation at different hierarchical levels when the host population is spatially structured. The aggregation of parasites can be partitioned between two levels, for instance among individual bird hosts within a nest and among nests. This approach can be useful to understand the exact mechanisms generating aggregated distributions, especially if aggregation occurs at one level but not at another.

6.1.2 PROBLEMS WITH THE MEASUREMENT OF AGGREGATION

Whatever the index of aggregation one chooses, it is likely to covary with other infection parameters. Typically, both prevalence and the mean number of parasites per host are positively correlated; both measures also

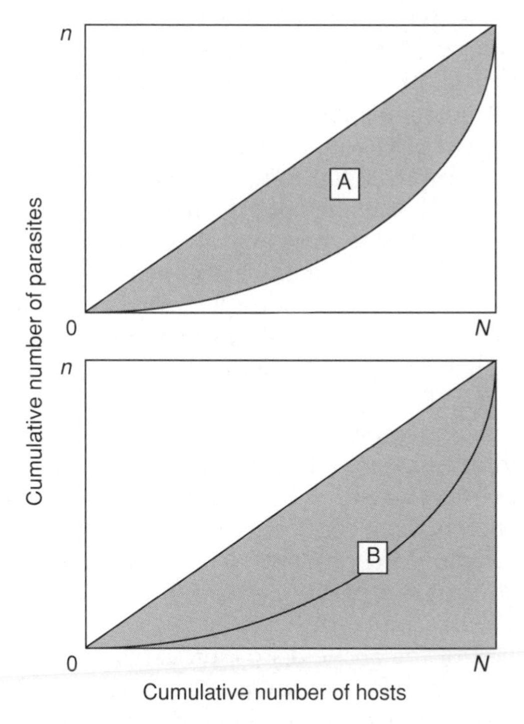

Figure 6.2 The relative discrepancy, *D*, between the observed distribution of parasites among hosts (curve) and their hypothetical uniform distribution (straight line) can be quantified as the ratio of area A to area B. The cumulative number of parasite individuals is plotted against the cumulative number of host individuals, with hosts ranked from the least to the most heavily infected prior to being summed up. (From Poulin, 1993.)

correlate negatively with aggregation as measured by *k* or *D* (Anderson, 1982; Poulin, 1993). Using the variance-to-mean ratio as a measure of aggregation gives different results (Dobson and Merenlender, 1991; Poulin, 1993; Shaw and Dobson, 1995). Some of these relationships are no doubt simple statistical properties of an aggregated distribution, and should not be interpreted as biological phenomena.

One variable which affects equally all measures of aggregation is host sample size (Ludwig and Reynolds, 1988; Poulin, 1993). True population aggregation levels are always underestimated when computed on small samples, producing a positive relationship between sample size and aggregation levels. This is purely a statistical artefact. This is clearly seen by generating a hypothetical population of hosts harbouring an aggregated parasite population, and drawing random samples of different sizes from this host population (Figure 6.3). At small sample sizes, estimates of both aggregation level and mean number of parasites per host have a high probability of underestimating (and a low probability of overestimating) true population values. This is an inevitable consequence of parasite aggregation: heavily-infected hosts are rare and unlikely to be included in small samples. Small samples are therefore not representative of the population. The accuracy of estimates of aggregation and of the mean number of parasites per host improves with further sampling, such that sample values increase asymptotically toward real population values as host sample size increases (Figure 6.3). Similar results are obtained independently of the measure used to quantify aggregation or mean number of parasites per host (Gregory and Woolhouse, 1993; Fulford, 1994; Poulin, 1996f). The measurement of prevalence is also plagued by the influence of host sample size (Gregory and Blackburn, 1991). In the sampling of hypothetical populations (Figure 6.3), the effect of sample size of estimates of prevalence is not as marked as its effect on the measurement of other parameters (Gregory and Woolhouse, 1993; Poulin, 1996f). The calculation of prevalence is affected not by the absence of heavily infected hosts but by the proportion of infected hosts, and this should not vary with sample size.

The main point of the above discussion is that host samples of small sizes are inappropriate for the estimation of aggregation. The more aggregated the parasite population, the more aggregation can be underestimated by relying on small host samples. Mathematical corrections for small sample sizes cannot be used as the true aggregation level is unknown. Therefore, whatever index is used, efforts should be made to obtain large, representative samples of hosts for computations of aggregation measures. This may be particularly relevant for studies of dynamics of parasite populations over time, in which host samples are taken at monthly intervals; because the availability of hosts or the ease with which they can be captured will vary with the seasons, the dynamic picture of aggregation thus obtained may often be a poor reflection of reality because of variable sample sizes.

6.2 CAUSES OF AGGREGATION

The almost universal occurrence of parasite aggregation suggests that similar processes may be acting to generate the same pattern in different host-parasite systems. Many biological forces operate to shape the distribution of parasites among hosts, some tending to increase aggregation and others tending to produce a more even distribution (Crofton, 1971; Anderson and Gordon, 1982). At any point in time, the degree of aggregation is determined by the balance of these forces.

The biological forces shaping aggregation include stochastic demographic mechanisms in both the host and parasite populations (Anderson

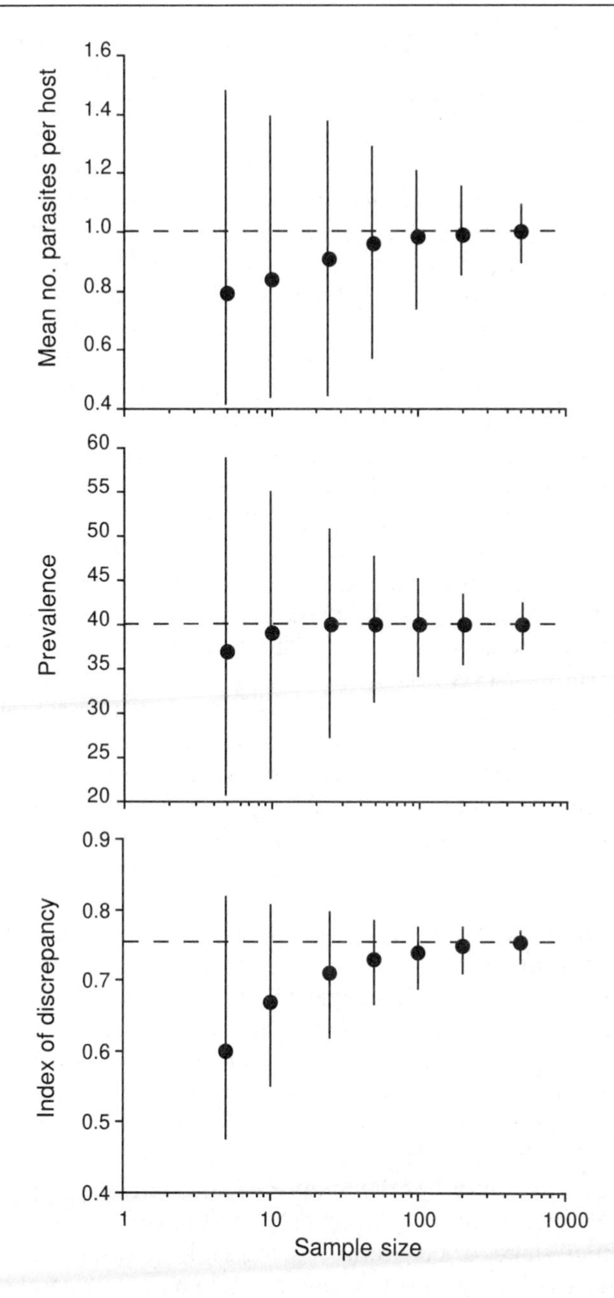

Figure 6.3 Relationship between host sample size and the mean number of parasites per host, the prevalence of infection, and the index of discrepancy. Data were obtained using bootstrap sampling of a hypothetical population of 1000 hosts harbouring 1000 parasites. Broken lines represent the true population values. For each host sample size, the mean and range of 25 simulated samples are presented. (From Poulin, 1996f.)

and Gordon, 1982). For instance, the direct reproduction of parasites within hosts would increase aggregation. At the same time, density-dependent parasite mortality and parasite-induced host mortality are two forces acting to decrease aggregation. These latter two processes prevent the formation of stable infrapopulations consisting of many parasites: such infrapopulations may occur temporarily but are invariably reduced in size or removed altogether from the population at large. Systems have been reported in which one or both of these processes are believed to act strongly and in which parasites are uniformly distributed among their hosts (Zervos, 1988a; Adlard and Lester, 1994; Donnelly and Reynolds, 1994). The copepod *Leposphilus labrei*, parasitic on wrasse off the Irish coast, provides an example: 1922 of the 1924 infected fish examined harboured a single copepod, and the other two harboured two copepods (Donnelly and Reynolds, 1994). Some form of infrapopulation control, perhaps host-mediated, prevents the existence of stable infrapopulations of two or more parasites. Systems with such strict infrapopulation regulation are exceptions, however.

Anderson and Gordon (1982) used Monte Carlo simulations to investigate the interactions among the processes affecting the rates of gain or loss of parasites by host individuals. They showed that heterogeneity in these rates among hosts was the major cause of aggregation when direct reproduction of parasites in hosts did not occur. If acquisition of parasites by hosts is a random process, so that the distribution of acquisition rate values follow a Poisson distribution, the resulting variance-to-mean ratio of parasite infrapopulation sizes will be greater than unity. In these conditions, parasite aggregation only becomes less pronounced when offset by processes such as a highly density-dependent parasite mortality rate (Anderson and Gordon, 1982).

Heterogeneity among hosts in rates of parasite acquisition can have at least two main origins. First, it may be due to heterogeneity in exposure to parasites among hosts. Hosts that are equally susceptible to infection may acquire different numbers of parasites simply because of the patchy distribution of parasites in space and time. For example, Keymer and Anderson (1979) showed that in experimental situations, aggregated distributions of infective stages in space accentuate the aggregated distribution of parasites in the host population (Figure 6.4). Janovy and Kutish (1988) modelled the effect of the temporal dispersion of parasite infective stages on the distribution of the parasites in their host. They found that when infective stages are continuously available, parasites are not particularly aggregated in their hosts. However, when the temporal dispersion of infective stages is highly heterogeneous, i.e. when infective stages are released in discrete waves, the parasites become highly aggregated in their hosts. The results of these simulations are in agreement with empirical measurements (Janovy and Kutish, 1988), and suggest that temporal as well as spatial heterogeneity in rates of exposure are enough to produce aggregated distributions.

The second cause of variability in rates of parasite acquisition results from heterogeneity among hosts in susceptibility to infection. Some hosts are better than others at preventing the initial establishment of parasites, or at eliminating them at a later stage following establishment. Other hosts may provide better habitats for parasites, regardless of immunity. In experimental infections, hosts of the same age, size and sex, given identical doses of infective stages, are often found harbouring different numbers of parasites. In such cases, variability in susceptibility to parasites reflects either or both differences in the ability of the host's defences (including the immune system) to fight parasitic infection, or differences in the quality of the hosts as habitats for parasites. There is considerable evidence that these differences have a genetic basis (Schad and Anderson, 1985). For instance, experiments with different genetic strains of hosts often demonstrate between-strains differences in susceptibility to infection (Wakelin, 1978, 1985). In addition, resistance to parasite infections is known to break down in hybrid zones between related host species, due to the mixing of genotypes (Sage *et al.*, 1986; Moulia *et al.*, 1993). Finally, it is possible to breed for parasite resistance in domestic

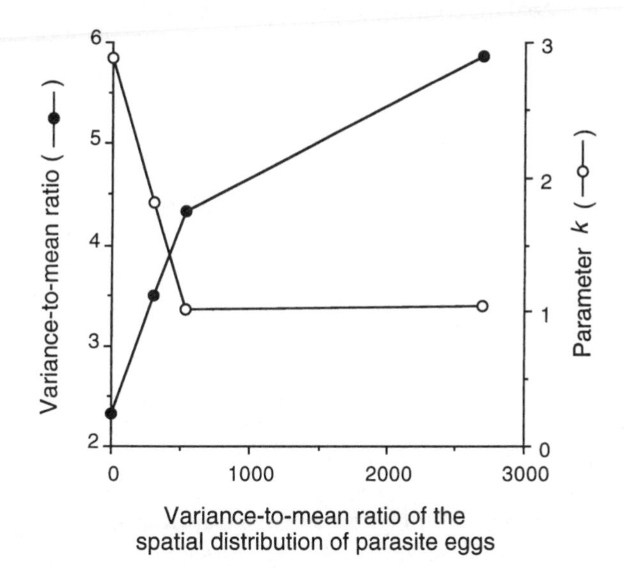

Figure 6.4 Relationship between the aggregation of parasites among their hosts (measured using both the variance-to-mean ratio and the parameter *k* of the negative binomial distribution) and the spatial distribution of parasite eggs. Data are from experiments in which equal numbers of beetles, *Tribolium confusum*, were exposed to identical numbers of eggs of the cestode *Hymenolepis diminuta*, for which it acts as an intermediate host; only the dispersion of eggs varied between treatments. (Modified from Keymer and Anderson, 1979.)

animals (Albers and Grey, 1986; Woolaston and Baker, 1996). All these lines of evidence indicate that some host individuals may be genetically predisposed to acquire large numbers of parasites relative to their conspecifics, even given equal exposure to infective stages.

Host genetics may be sufficient to produce an aggregated parasite distribution in the absence of all other ecological variables. Wassom *et al.* (1986) found that the immune response that expels the cestode *Hymenolepis citelli* from the small intestine of its definitive rodent host is under genetic control. Resistance is conferred by a dominant allele at a single gene locus. Patent infections with the cestode occur only in individuals that are homozygous for the recessive allele: these hosts are incapable of rejecting the cestode before it matures. Wassom *et al.* (1986) suggest that the aggregated distribution of *H. citelli* in natural host populations is due primarily to the high frequency of resistant host genotypes. The simple one-locus genetic resistance in this host-parasite system may not be representative of the majority of systems. Indeed, in a related system (same cestode in a congeneric host), the genetically based expulsion of the parasite plays only a minor role in generating an aggregated parasite distribution (Munger *et al.*, 1989). Immunity may be generally unimportant in the field as a causative agent of parasite aggregation (see Quinnell *et al.*, 1990) in contrast with variability in other host traits.

In fact, the heterogeneity among hosts in susceptibility to parasites responsible for parasite aggregation will usually result from a combination of factors. It may not be entirely innate, and some heterogeneity may be acquired. The best example is acquired immune resistance, which is a function of both innate mechanisms and environmental influences such as exposure to parasites at a young age. Often heterogeneity in susceptibility and thus parasite aggregation will be the product of interactions between innate and acquired resistance in the host population (Tanguay and Scott, 1992). Acquired heterogeneity in susceptibility can take other forms, too. In many ectoparasitic arthropods, pheromones released by established parasites are believed to attract new parasites to infected hosts (e.g. Blower and Roughgarden, 1989; Norval *et al.*, 1989). In others, parasite-induced changes in host behaviour seem to make infected hosts more easily detected by parasites and more vulnerable to further infections (Poulin *et al.*, 1991). In these systems, early exposure to and infection by parasites could turn some hosts into magnets attracting further parasites at a higher rate than uninfected conspecifics. Acting in isolation, this process could easily and rapidly generate aggregated parasite distributions.

All the mechanisms influencing parasite distributions discussed above will vary in strength among host-parasite systems. There may be properties of hosts or parasites that affect the degree of aggregation. In comparative analyses of published values of parasite aggregation, for instance, there is a tendency for lower levels of aggregation to be observed in intermediate than in definitive hosts, and for ectoparasites to be more aggregated than endoparasites (Dobson and Merenlender, 1991). Also, in systems involving vertebrate hosts, parasites that enter the host passively when ingested accidentally, such as trichostrongylid and oxyurid nematodes, tend to be much more aggregated than parasites such as acanthocephalans that enter the host when the latter ingests an infected intermediate host (Shaw and Dobson, 1995).

Among parasites with similar life cycles and modes of transmission, the spatial distribution of hosts may explain variation in levels of aggregation. In experimental situations, mobile infective larvae achieve a higher prevalence of infection and/or lower levels of aggregation when target hosts are clumped than when hosts are randomly or evenly distributed (Blower and Roughgarden, 1989; McCarthy, 1990). Along the same lines, Rózsa *et al.* (1996) and Rékási *et al.* (1997) found that avian lice species parasitizing colonial host species are less aggregated than related lice species parasitizing territorial hosts. Therefore, aggregation of hosts may weaken the aggregation of some parasites, by facilitating transmission and minimizing differences among host individuals in levels of exposure to infective stages. Many other host characteristics can influence parasite aggregation. For example, Møller (1996) suggested that parasite aggregation should vary positively with the intensity of sexual selection in the host species. However, this relationship, as well as the effects of other host and parasite traits, remains to be tested empirically.

6.3 CONSEQUENCES OF AGGREGATION

Aggregation of individuals among hosts is the rule in metazoan parasite populations. Clearly this pattern of distribution can influence the evolutionary biology of parasites. We may expect the frequency of interactions between conspecific parasites to be determined by the level of aggregation. Different individual parasites may experience highly different levels of competition for resources, depending on the size of the infrapopulation in which they occur. Many parasites will be in small infrapopulations and will not suffer from competition, whereas many other parasites will belong to fewer but larger infrapopulations and will experience intense competition for food or space (Figure 6.1). The frequency of encounter between mates is also likely to be highly uneven among parasites because of aggregation, as are the individual reproductive success of parasites and the fate of entire infrapopulations. This section will discuss some of the consequences of parasite aggregation on three selected aspects of parasite evolutionary ecology.

6.3.1 POPULATION STRUCTURE AND GENETICS

The most obvious effect of parasite aggregation on parasite biology is that the intensity of intraspecific competition for space or nutrients will not be equal for all individuals in a population, but will instead be proportional to infrapopulation size. Other living conditions in the host can also vary among infrapopulations as a consequence of the variability in infrapopulation size. For these and other reasons, parasites in infrapopulations of large sizes experience a 'crowding effect' (Read, 1951), and both growth and fecundity show density dependence in many helminth taxa (Keymer *et al.*, 1983; Keymer and Slater, 1987; Jones *et al.*, 1989; Quinnell *et al.*, 1990; Shostak and Scott, 1993). The shape of the relationship between parasite growth or fecundity and infrapopulation size determines the strength of the density dependence; but for most realistic functions we can expect that aggregation will lower average parasite fitness (Jaenike, 1996c).

Although average fitness is lower in large infrapopulations, the variance in fitness is typically large (Dobson, 1986). At both levels, of the various infrapopulations and the entire population, this may result in gross inequalities in size or fecundity among individuals (Dobson, 1986). The adult parasite population in the definitive hosts may consist of several small individuals with low fecundity and a few large individuals with high fecundity, all sexually mature and producing eggs at a rate proportional to their size. This phenotypic plasticity in growth and development may have been favoured over evolutionary time because the aggregated distribution of parasites prevents all individuals from reaching their potential adult size. It is not yet clear whether the minority of parasites that attain large sizes are individuals with superior genotypes, or whether adult size results solely from chance events following infection such as arriving first and securing a good attachment site.

Dobson (1986) proposed the use of Lorenz curves and Gini coefficients to measure the degree of inequality in size or fecundity among conspecific parasites in a population (Figure 6.5). This approach was also the inspiration behind the index of discrepancy used to quantify aggregation levels (see section 6.1). Dobson (1986) suggested that Gini coefficients computed using the distribution of total biomass among individuals could indicate to what extent a large share of the reproduction in a population is undertaken by only a few individuals. Later studies in host-parasite systems where measures of parasite fecundity can be obtained directly have shown that inequalities in egg production are more pronounced than inequalities in body size (Shostak and Dick, 1987; Szalai and Dick, 1989). Typically, in highly aggregated parasite populations, only a few worms may be responsible for most of the total egg output, such that genetic contributions to the next generation are highly unequal among individuals.

In population genetics, it is the effective population size, or the number of individuals contributing genes to the next generation, and not the actual population size that determines the relative amount of genetic drift (Hartl and Clark, 1989). In aggregated parasite populations experiencing

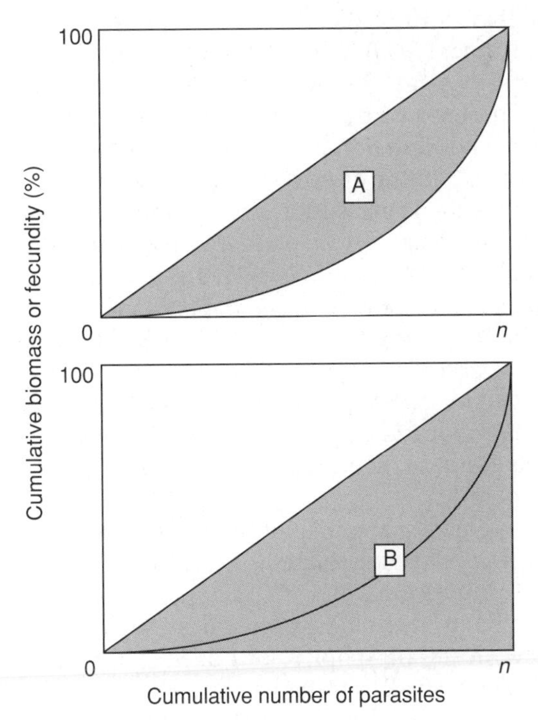

Figure 6.5 The Gini coefficient, used to quantify inequalities among parasites in size or egg production, is simply the relative discrepancy between the observed distribution of biomass or fecundity among parasites (curve) and their hypothetical uniform distribution (straight line). It can be quantified as the ratio of area A to area B. The cumulative biomass or fecundity is plotted against the cumulative number of parasite individuals, with parasites ranked from the smallest or least fecund to the largest or most fecund prior to being summed up. (From Dobson, 1986.)

inequalities in reproduction, the effective population size may be closer to the number of infected hosts than it is to the total number of parasites, because only very few parasites per host achieve high egg outputs regardless of infrapopulation sizes (Dobson, 1986). The smaller the effective population size, the greater the likelihood that the allele frequencies and the genetic heterogeneity characteristic of a parasite population at a point in time will not be transmitted to the next generation. At small effective population sizes, the heterozygosity of the population may be reduced because of inbreeding, and random changes in allele frequencies caused by stochastic events become more likely (Hartl and Clark, 1989; Nadler, 1995). This effect can be further enhanced by skewed sex ratios, which are characteristic of many dioecious parasite taxa (see section 6.3.2).

Studies of genetic variability within and among parasite populations are still few, but recent developments in molecular techniques should pave the way for future studies (Nadler, 1995). In a survey of the available evidence, Nadler (1990) found that levels of genetic variation in populations of helminth parasites were a little lower than those of free-living invertebrates. The level of heterozygosity, or the average proportion of heterozygous loci per individual, was 0.07 for parasites and 0.11 for free-living invertebrates (0.10 if insects are excluded; Nevo, 1978). However, when genetic variation is measured as the proportion of polymorphic loci among the total number of loci surveyed, parasite populations appear much less variable than those of free-living invertebrates. The average level of polymorphism among parasite populations is 0.23 (see Figure 6.6) whereas that of free-living invertebrates is 0.40 (the result is the same when insects are excluded; Nevo, 1978). The very low levels of genetic variation observed in some helminth populations (Figure 6.6) may be due to the cumulative reduction in genetic variability resulting from a small effective population size maintained over several generations (Bullini *et al.*, 1986; Nadler, 1995).

The effects of parasite aggregation and the associated inequalities in reproduction on effective population size and on the genetic variation within parasite populations may be minimal for certain parasites. For example, directly transmitted nematodes such as *Ostertagia* and *Haemonchus* often form infrapopulations of several thousand individuals in their hosts. Their small size limits the effects of crowding on their reproductive output, and inequalities in fecundity are not likely. Typically, these parasites show high levels of genetic variation within

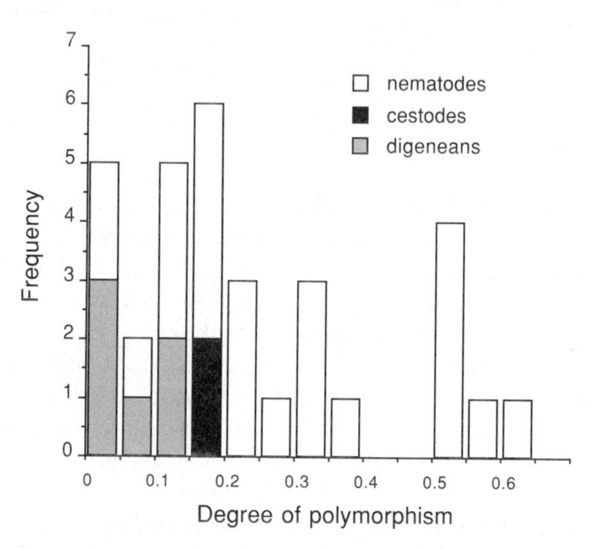

Figure 6.6 Frequency distribution of genetic polymorphism levels among 32 parasite populations, representing 23 helminth species. Polymorphism is measured as the proportion of polymorphic loci among those surveyed. (Data from Nadler, 1990.)

populations (Blouin *et al.*, 1992, 1995). In contrast, *Ascaris suum*, a nematode with a similar life cycle and also parasitic in livestock, displays high levels of within-population homozygosity and high inbreeding coefficients (Nadler *et al.*, 1995). The low genetic variation among *A. suum* individuals is most probably due to their much larger body size, and crowding may therefore have a greater negative effect on the evenness of reproductive output among individuals.

We may expect an even greater tendency toward low levels of genetic variation in parasites with complex life cycles involving asexual reproduction in an intermediate host. In the digenean *Fascioloides magna*, for instance, asexual reproduction in the snail intermediate host may result in a patchy distribution of metacercariae, with each patch consisting of metacercariae derived from a single parasite egg. Mulvey *et al.* (1991) found that deer, the definitive host of the parasite, tended to harbour infrapopulations of adult worms representing a single clone. The minimal genetic diversity observed in adult infrapopulations of this digenean may not be characteristic of all digeneans. Recently, Minchella *et al.* (1995) found that over half of the snail intermediate hosts of *Schistosoma mansoni* in their sample harboured multiple parasite genotypes, i.e. they released cercariae descended from different eggs. The high genetic diversity of *S. mansoni* infrapopulations is also apparent in rats, the natural definitive hosts of the parasite (Barral *et al.*, 1996). Lymbery and Thompson (1989) found a similar pattern in the cestode *Echinococcus granulosus*, which also multiplies asexually in its intermediate host. Individual intermediate hosts contain *E. granulosus* cysts derived from genetically different embryos, such that genetic heterogeneity persists in intermediate hosts despite the asexual multiplication of cestode larvae. In any event, the reduction in the genetic diversity of adult infrapopulations resulting from asexual multiplication in intermediate hosts, be it great or small, can only compound the effect of aggregation on the likelihood of genetic drift and place the gene pool of the parasite population at the mercy of stochastic events.

6.3.2 SEX RATIO

The highly fragmented nature of parasite populations resulting from aggregation, and the associated low levels of genetic variation within infrapopulations, can favour sex ratios departing from unity. Equal investments in offspring of the two sexes should be selected under a wide range of conditions, but biased sex ratios can also evolve (Charnov, 1982; Godfray and Werren, 1996). Factors that can push the population sex ratio away from unity over evolutionary time include the probability of mating and the likelihood of inbreeding. Because of the aggregation of parasite individuals among hosts, biased sex ratios are to be expected in parasite populations.

May and Woolhouse (1993) and Morand *et al.* (1993) have modelled the consequences of aggregation and the type of mating system on the probability of mating in parasites. In monogamous mating systems, the pairing probability of female parasites will be higher when sex ratios are male-biased, especially when the population is highly aggregated. In polygamous mating systems, female-biased sex ratios can increase the probability of mating, particularly when aggregation is not pronounced. The results of these mathematical models suggest that the distribution of parasites can exert a selective pressure on the sex ratio of the parasite population.

In accordance with the models, sex ratios are almost invariably male-biased in the monogamous schistosomes (Mitchell *et al.*, 1990; Morand *et al.*, 1993). The bias appears to be a true adaptive phenomenon and not merely a consequence of differential mortality between the sexes, as the sex ratio of cercariae emerging from snails is also male-biased (Mitchell *et al.*, 1990). Schistosome sex ratios in the definitive host show density dependence, and tend toward unity as infrapopulation size increases (Morand *et al.*, 1993). Thus the population structure of the parasite can also have a proximate effect on sex ratios and interact with genetically programmed patterns of male and female offspring production.

In the polygamous nematodes (Roche and Patrzek, 1966; Guyatt and Bundy, 1993) and acanthocephalans (Crompton, 1985), female-biased sex ratios have been documented repeatedly over the years and are much more frequently observed than male-biased ratios (Figure 6.7). In these worms, the sex ratio of juveniles is closer to unity, and the biases observed in adult populations are caused by mechanisms acting between hatching and adulthood. The key mechanism appears to be differential mortality between the sexes, with females surviving longer than males following infection (Poulin, 1997d). Seasonal fluctuations in sex ratio in natural populations and temporal changes in sex ratio in experimental infections both provide evidence for sex-dependent mortality rates. However, sex ratios of nematodes are also density-dependent (Tingley and Anderson, 1986; Stien *et al.*, 1996), and population structure could influence sex ratios and lead to adaptive adjustments in the proportions of males and females in parasite populations. A comparative analysis provided little evidence supporting a role for aggregation or population structure in the evolution of parasite sex ratios in nature, although a similar analysis using data from experimental infections indicated that at high intensities of infection, sex ratios are less female-biased (Poulin, 1997d).

Protozoans parasitic in the blood of vertebrates also display female-biased sex ratios (Schall, 1989; Read *et al.*, 1995; Shutler *et al.*, 1995). These parasites differ from helminths in that they multiply directly within their host; in this situation, prevalence of infection becomes a better indicator of the probability of multiple infections (i.e. the frequency of infrapopula-

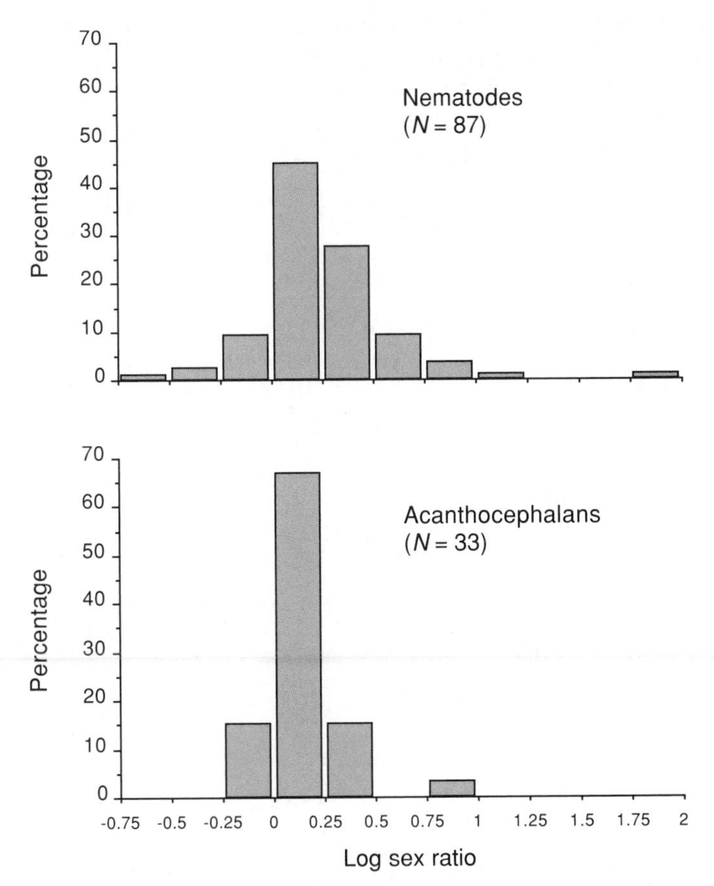

Figure 6.7 Frequency distribution of female-to-male sex ratios among adult populations of parasitic nematodes and acanthocephalans. Values smaller than zero indicate male-biased ratios, and values greater than zero (i.e. the majority of values) indicate female-biased ratios. (From Poulin, 1997d.)

tions consisting of mixed parasite genotypes) than aggregation. Usually, as either prevalence or mean infrapopulation size increases, the frequency of mixed infections increases also. As female-biased sex ratios are favoured when inbreeding is likely, the degree of female bias should vary with prevalence or with mean infrapopulation size (Read *et al.*, 1992). Comparative analyses across populations of two genera of blood parasitic protozoans yielded contrasting results, with only one study supporting the prediction (Figure 6.8). Clearly, parasite sex ratios are influenced by several factors, some acting before infection and others acting after. There is mounting evidence that one of the consequences of aggregation of parasites at the adult stage could be a shift in the population sex ratio away from unity; however, other forces acting in the other direction can make the role of aggregation difficult to distinguish. Comparative analyses in other taxa of parasites, such as arthropods in which mating systems and sex ratios vary within species (Rózsa, 1997) and across species (Anstensrud, 1990; Heuch and Schram, 1996; Rózsa *et al.*, 1996), may provide further insight into this problem.

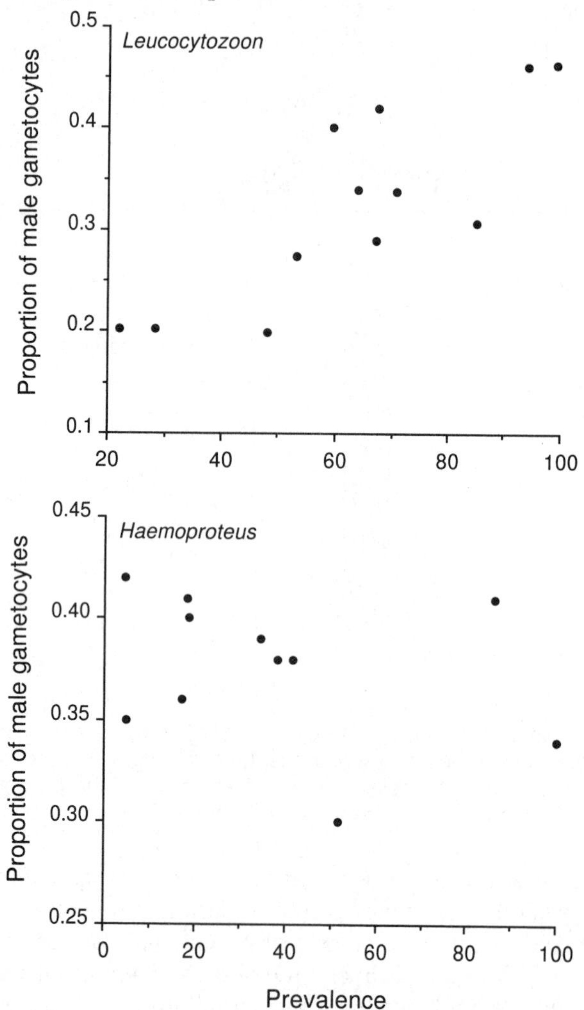

Figure 6.8 Relationship between prevalence of infection and the proportion of male gametocytes in the blood of infected hosts in two genera of protozoans parasitic in birds. The proportion of males tends toward 0.5 as prevalence increases among 12 populations/species of the genus *Leucocytozoon*, but not among 11 populations/species of the genus *Haemoproteus*. (Modified from Read *et al.*, 1995; Shutler *et al.*, 1995.)

6.3.3 MACROEVOLUTIONARY PHENOMENA

The association between spatial distribution and the probability of extinction is well recognized in ecology and biogeography (Hanski, 1982; Hengeveld, 1992). In free-living invertebrates, the probability of regional extinction decreases as the number of localities occupied within a region increases (Hanski, 1982). For parasite populations, the probability of extinction may well decrease with the relative number of hosts occupied, or with the prevalence of infection.

This can best be seen by considering hypothetical extreme cases. When aggregation of parasites is at its maximum and prevalence at its minimum, i.e. all parasites are in a single host, the parasite population is in jeopardy. Because the parasite-induced host mortality is usually dependent on the number of parasites per host (Anderson and Gordon, 1982), the infected host is likely to die and take the entire parasite population with it. At the other extreme, when prevalence is maximum and there is no aggregation, i.e. the parasites are evenly distributed among all available hosts, the only way the parasite population can become extinct is if the entire host population goes extinct. Actual parasite populations show intermediate levels of aggregation, resulting in a few large infrapopulations having a high risk of extinction and several small infrapopulations having a much lower risk of extinction (Price, 1980). The entire parasite population only becomes at risk when aggregation levels are high for long periods.

A high probability of extinction for highly aggregated parasites in founder host populations could explain some of the incongruence between the phylogeny of parasites and that of their hosts caused by missing parasite branches (Paterson and Gray, 1996). Founder host populations often have fewer parasite species than their population of origin. For instance, the possum *Trichosurus vulpecula*, introduced to New Zealand from Australia, seems to have a poorer parasite fauna in New Zealand than in its country of origin (Stankiewicz *et al.*, 1996). A total of only 200–300 individuals, from several sites in Australia, were released in many localities throughout New Zealand. As some of the parasite species missing in New Zealand have direct life cycles, and suitable intermediate hosts are available for others, founder effects resulting from parasite aggregation may explain the discrepancy between the New Zealand and Australian parasite faunas.

There is no empirical evidence supporting a role for aggregation in parasite extinction, and it may prove difficult to obtain any. It may be possible to contrast the regional occurrence of parasite taxa with different aggregation levels. For instance, we might predict that highly aggregated parasites will be present in fewer host populations, such as populations of a fish species in different lakes, than weakly aggregated parasites with similar life cycles. In any event, with or without these data, the effect of parasite distribution among hosts on the risk of extinction is probably very strong.

6.4 SUMMARY

The host population represents a collection of resource patches among which parasites are unevenly distributed. The aggregation of parasites among hosts produces many small infrapopulations and a few large ones, so that the number of conspecifics sharing the same host is extremely variable. The selective pressures exerted on parasites by the very nature of their distribution may have contributed, among other features, to the plastic growth and developmental schedules and biased sex ratios exhibited by many parasites.

It is often difficult to distinguish between the causes and consequences of aggregation among the various genetic, individual and population processes related to parasite distribution. Nevertheless, almost all parasites are aggregated to some degree among their hosts, to the point that Crofton (1971) proposed that this typical population distribution should become part of the definition of what is a parasite. One is left wondering whether the aggregation level of a parasite population is the product of selection or simply a non-adaptive consequence of various population processes. Shaw and Dobson (1995) believe that the intermediate levels of aggregation shown by most parasite populations are the outcome of selection. Reduced mating opportunities when aggregation is low, and increased competition and extinction when aggregation is high, would have favoured the aggregation levels observed today. Adaptive or not, the clumped dispersion of parasites has permeated all aspects of parasite ecology and evolution. Nonetheless, its influence on the biology of parasite individuals and populations has not been considered in many recent studies. This failure will need to be corrected if parasite evolution is to be placed in the proper context of highly variable infrapopulation sizes.

Parasite population dynamics 7

Parasite populations vary in size over both short and long time scales and are affected by both biotic and abiotic factors. Some of these factors produce changes in parasite numbers, whereas others act to reduce the amplitude of fluctuations around an equilibrium population size. The study of parasite population dynamics has been tightly linked with epidemiology, which is concerned with the spread of disease in host populations (Anderson and May, 1991; Anderson, 1993). The theoretical framework developed in epidemiology is proving useful for the interpretation of field and experimental observations, and has played a key role in motivating parasitologists to obtain accurate measurements of population parameters important for the study of parasite transmission.

This chapter will begin by reviewing some mathematical models of parasite population dynamics, and will progress toward a discussion of selected case studies from the field. The emphasis will be on identifying the key parameters influencing parasite numbers over time, and on how these may have been shaped by natural selection.

7.1 MODELS OF PARASITE POPULATION DYNAMICS

In the late 1970s, Anderson and May produced the first models of host and parasite population dynamics (Anderson, 1978; Anderson and May, 1978, 1979; May and Anderson, 1978, 1979). These have served as a foundation for subsequent models, which are all essentially modifications of the original Anderson and May models adapted for more specific circumstances. In the models, the temporal variation in host or parasite numbers is determined by some key instantaneous rates (Table 7.1). Thus the size of the parasite population will change as a function of the overall rate at which individuals are lost and the overall rate at which new individuals are recruited. The way in which the models can be used is best illustrated by applying them to specific life cycles.

Table 7.1 Key parameters used in models of parasite population dynamics

Parameter	Description
P	Size of adult parasite population in definitive host, or in only host if life cycle is direct
P'	Size of juvenile parasite population in intermediate host
W	Size of parasite egg population in external environment
H	Size of definitive host population
N	Size of intermediate host population
μ	Instantaneous death rate of adult parasites within host, due to natural or immunological causes (/adult parasite/*t*)
b	Instantaneous death rate of definitive host, where mortalities are due to causes other than parasites (/definitive host/*t*)
α	Instantaneous death rate of the definitive host, where mortalities are induced by parasites (/definitive host/*t*)
λ	Instantaneous rate of production of infective stages, e.g. eggs, cysts (/adult parasite/*t*)
p	Instantaneous rate of predation of definitive hosts on intermediate hosts (/(definitive hosts)(intermediate hosts)/*t*)
e	Instantaneous death rate of parasite eggs (/egg/*t*)
n	Instantaneous rate of ingestion of parasite eggs by intermediate hosts (/intermediate host/*t*)
v	Instantaneous death rate of juvenile parasites within host, due to natural or immunological causes (/juvenile parasite/*t*)
d	Instantaneous death rate of intermediate host, where mortalities are due to causes other than parasites (/intermediate host/*t*)
H_0	Transmission efficiency constant, which varies inversely with the proportion of infective stages that successfully establish in hosts
k	Aggregation parameter of the negative binomial distribution (see Chapter 6)

The first example examines the dynamics of an aggregated population of a helminth parasite with a direct life cycle. Adult parasites inside or on a host produce infective stages that leave the host to infect other hosts. Many nematodes and monogeneans are examples of this type of cycle. Some simplifying assumptions can be made at first, such as supposing that the parasite-induced host death rate is linearly proportional to the number of parasites per host, and that parasite mortality and fecundity within hosts are not density-dependent. If we further assume that transmission is virtually instantaneous without development in the external environment, we can express changes in the parasite population as a single differential equation (Anderson and May, 1978; May and Anderson, 1979):

$$\Delta P/\Delta t = (\lambda HP/H_0+H) - P(\mu+b+\alpha) - \alpha(k+1)P^2/kH$$

where changes in parasite numbers are determined by the numbers of new parasites joining the population minus the numbers of parasites dying of various causes. All symbols are defined in Table 7.1. The flow of individuals in the model can be visualized in Figure 7.1.

Changes in the size of the parasite population can also be examined mathematically using the expression for the basic transmission rate of the parasite, R_0 (Anderson and May, 1991). In Chapter 5, R_0 was used as a measure of the parasite's lifetime reproductive success and was linked with the evolution of virulence. It is more commonly used to calculate the threshold host population size below which the parasite population cannot survive (i.e. host population size below which R_0 is negative) or the present dynamic state of the parasite population (stable if $R_0 = 1$, increasing if $R_0 > 1$).

The above equation succinctly expresses the dynamic nature of the parasite population. It is based on simplifying assumptions that can be replaced with more realistic ones. By doing this and by analysing the influence of each parameter in the model, it is possible to determine the relative contribution of each variable to parasite population dynamics. For instance, in situations where the rate of parasite-induced host death increases, say, exponentially rather than linearly with the number of parasites per host, the stability of both host and parasite populations is likely to be enhanced. Similarly, density-dependent effects on parasite survival

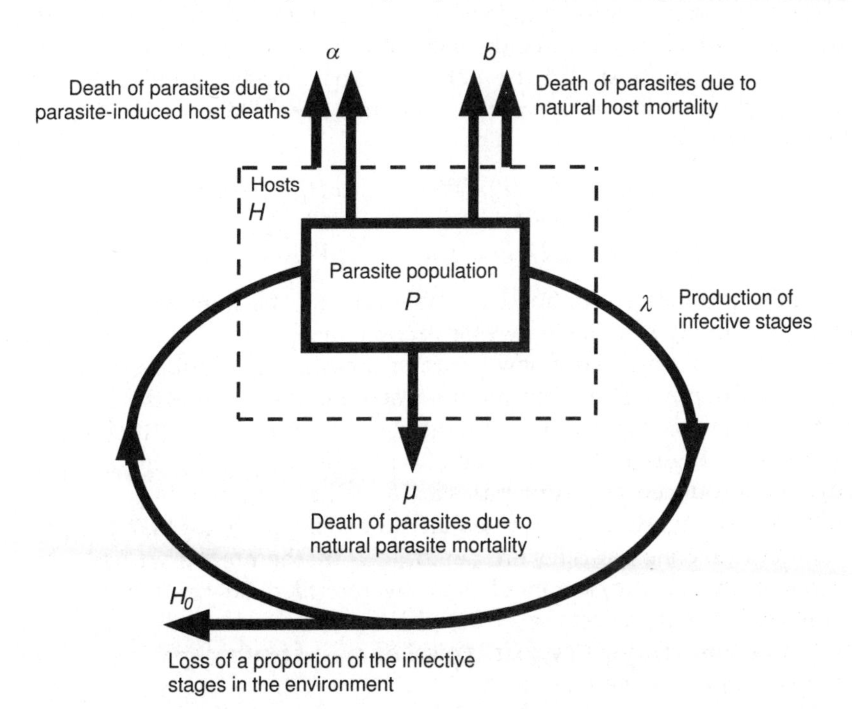

Figure 7.1 Summary of gains and losses of individuals in a parasite population with a direct life cycle. The parameters are defined in greater detail in Table 7.1.

or fecundity within infrapopulations will also contribute to the overall stability of both host and parasite populations (Anderson and May, 1978). The following section will review the evidence for density dependence in parasite infrapopulations.

Models such as the one described above are tools that can be put to several uses. The basic model can be adapted for slightly different life cycles and can be used to examine specific systems. For instance, models can facilitate the interpretation of observed population changes (Scott and Anderson, 1984; Keymer and Hiorns, 1986; Morand, 1993) or help to identify key population parameters around which control strategies can be designed (Smith and Grenfell, 1985).

The example of a parasite with a direct life cycle and instantaneous transmission without any time spent outside the host represents the simplest possible scenario, one in which the entire parasite population is always found in a single physical habitat, the host population. When complex life cycles are involved, the parasite population becomes compartmentalized into different hosts, with the possibility of one or more free-living stages as well. In a two-host life cycle where transmission from the intermediate host to the definitive host is by predation and where eggs released by adult parasites must be ingested by the intermediate host, the parasite population consists of three types of individuals (eggs, juveniles and adults) in three distinct habitats (the external environment, the intermediate host and the definitive host). Changes in these three parts of the population can also be modelled (from Dobson and Keymer, 1985; Dobson, 1988):

$$\Delta P/\Delta t = pHN[P'/N + (P'/N)^2] - P(b + \mu) - \alpha H[P/H + (P/H)^2(k + 1)/k]$$
$$\Delta W/\Delta t = \lambda P - eW - nWN$$
$$\Delta P'/\Delta t = nWN - P'(v + d) - pHN[P'/N + (P'/N)^2]$$

Again, symbols are defined in Table 7.1 and the flow of parasites is illustrated in Figure 7.2. Changes in the population of adult parasites are determined by how many new parasites are acquired through predation by the definitive host on infected intermediate hosts, minus the numbers of adult parasites dying either of natural causes or because of their effects on hosts. Changes in the egg population are defined by the rate at which eggs are produced by adult parasites and the rates at which they are either lost or ingested by intermediate hosts. Finally, changes in the population of juvenile parasites are determined by the number of new ones recruited through ingestion of eggs by intermediate hosts minus the numbers lost either because of natural mortality or transmission to the definitive host. This life cycle is typical of most acanthocephalans and of many cestodes and nematodes.

Again, analysis of these equations suggests that density-dependent constraints on adult survival and reproduction can regulate the whole

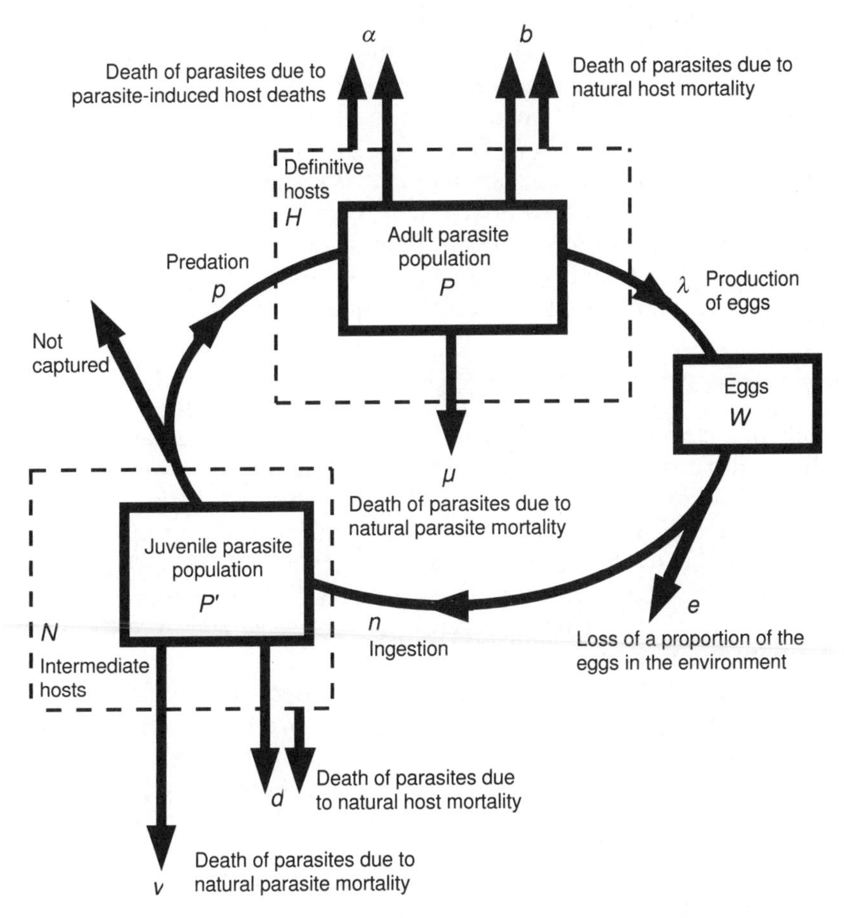

Figure 7.2 Summary of gains and losses of individual adults, juveniles and eggs in a parasite population with a two-host life cycle. The parameters are defined in greater detail in Table 7.1.

parasite population and lead to long-term stability. The same modelling approach, or a slightly modified one, has also been used to investigate the consequences of either parasite manipulation of the intermediate host's susceptibility to predation by the definitive host (Dobson and Keymer, 1985; Dobson, 1988) or the addition of a new intermediate host to the cycle (Morand *et al.*, 1995).

The above models apparently assign little importance to abiotic factors, which can cause abrupt changes in population size. There is no doubt that abiotic phenomena cause fluctuations in parasite populations (see Esch and Fernández, 1993). Physicochemical variables will determine the survivorship of eggs or other infective stages in the external environment. The growth and fecundity of adult parasites in ectothermic hosts can also be influenced by the external temperature. These density-independent processes usually do not, however, account for the long-term stability of parasite populations. Some would argue that parasite populations are generally unstable and not subject to any intrinsic regulation (Price, 1980). Mathematical models assume that parasite populations are regulated by density-dependent phenomena, which control the growth of the population. There is enough evidence in favour of population regulation and density dependence (see Keymer, 1982, and section 7.2) to conclude that the models are based on sound biological assumptions.

7.2 DENSITY-DEPENDENT REGULATION

Mathematical models focus on the dynamics of the whole parasite population (all parasite individuals in all hosts). In practice, though, any density-dependent regulating process leading to changes in population abundance will result from mechanisms acting within infrapopulations (all individuals within one host), which form interacting groups within the otherwise fragmented population. Thus overall population regulation is achieved by the sum of events taking place within the various infrapopulations.

The role of aggregation highlighted by mathematical models becomes clear. In a highly aggregated parasite population, there will be a few large infrapopulations (containing a large proportion, if not most, of the parasites in the population) in which density-dependent processes will act strongly. From the perspective of the host, regulation of parasite numbers is achieved through mechanisms acting in only a fraction of the infected hosts. In another parasite population of the same size but distributed more evenly, there may be no infrapopulations of a size greater than the threshold size above which regulating processes become important. This threshold size is not meant as some fixed infrapopulation size; obviously, what is considered crowded in one host may be considered sparse in another. Characteristics of the host as well as the presence of other parasite species will influence the infrapopulation size at which density-dependent processes become important. Nevertheless, parasite aggregation is the main force determining whether there will be enough dense infrapopulations for population regulation to occur. As aggregation of parasites is the rule (see Chapter 6), we may expect density-dependent regulation to be common in parasite populations.

Infrapopulations are short-lived, their maximum longevity being equal to the life span of the host. During their existence, infrapopulations undergo a constant turnover of individuals. New individuals are gained by recruitment of infective stages and old ones are lost through mortality. In some parasites, new individuals can be gained through the direct reproduction of parasites inside the host, and in rare species parasites can be lost or gained through emigration and immigration; these possibilities are not considered any further here. Density-dependent effects on survival, growth and fecundity of parasites (e.g. parameters μ, λ and v in the models) in infrapopulations will regulate the overall parasite population abundance. Increases in mortality at high parasite density will regulate the size of given infrapopulations, and decreases in growth and fecundity at high parasite density will contribute to regulation by influencing the availability of infective stages for all infrapopulations. The relationship between density (or the number of parasites per infrapopulation) and the magnitude of changes in survival, growth or fecundity may be roughly linear. But it may also assume other shapes. If, for instance, a threshold density must be reached before mortality increases or fecundity decreases, then the relationship may approximate an exponential function.

The mechanisms responsible for density-dependent regulation of parasite numbers are numerous, but are rarely identified in specific studies. Here, four possibilities are considered: exploitation competition; interference competition, in which individuals secrete chemicals that harm their competitors; host-mediated restriction; and parasite-induced host mortality. These may operate simultaneously, which explains why it is proving difficult to elucidate precisely which mechanism is operating in a host-parasite system.

The presence of one parasite in a host and its exploitation of the host means that there will be fewer resources available for other parasites of the same species. However, as long as individual parasites obtain sufficient resources, the effect of exploitation competition will be negligible. Only above a certain density, or a certain infrapopulation size, will the depletion of resources by conspecifics affect any given parasite. For instance, attachment space in the gut of the definitive host becomes a limiting resource for some acanthocephalans beyond a certain infrapopulation size (Uznanski and Nickol, 1982; Brown, 1986). The initial establishment and subsequent survival of worms show a strong dependence on infrapopulation size (Figure 7.3).

In other systems, both space and nutrients may be in short supply in dense infrapopulations. Oxyurid nematodes living in the gut of insects show reductions in *per capita* egg production in dense infrapopulations (Figure 7.4). These nematodes provide good examples of the difficulty involved in distinguishing between exploitation competition and interference competition. The mechanism behind the density-dependent fecundity of females (Figure 7.4) is likely to be competition for resources, but could also result from interference competition (Adamson and Noble, 1993). There also appears to be interference competition between the sexes (Zervos, 1988a, b; Morand and Rivault, 1992). Females may secrete a chemical that regulates the number of males per infrapopulation: the first male joining the infrapopulation survives but further males are not allowed to establish (Zervos, 1988a, b). Similarly, the 'crowding' effect commonly reported in dense cestode infrapopulations (see Chapter 6) may be the product both of competition for nutrients or other resources (Keymer *et al.*, 1983), and of competition through chemical interference

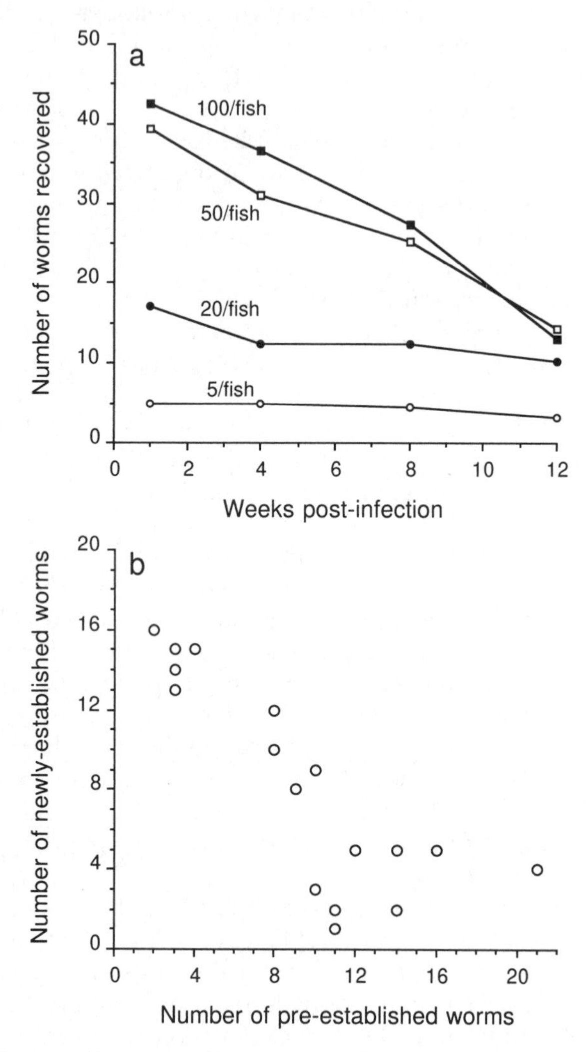

Figure 7.3 Evidence for density-dependent establishment and survival of the acanthocephalan *Pomphorhynchus* laevis in experimentally infected rainbow trout, *Oncorhynchus mykiss*. (a) Recovery of worms over time in four treatments of fish hosts given different initial numbers of infective cystacanths. Each point is the mean of 4–14 fish. (b) Establishment of new worms in fish harbouring various numbers of pre-established, 12-week-old worms. All fish were given 20 cystacanths in the second infection; each point represents a different fish. (Modified from Brown, 1986.)

(Zavras and Roberts, 1985). Whatever the nature of the mechanisms, the population consequences are the same.

Host-mediated restriction of parasite numbers in infrapopulations is another potential density-dependent mechanism. For instance, when rodents are first exposed to a dose of schistosome cercariae and then re-exposed to an identical dose a few weeks later, it appears that many more

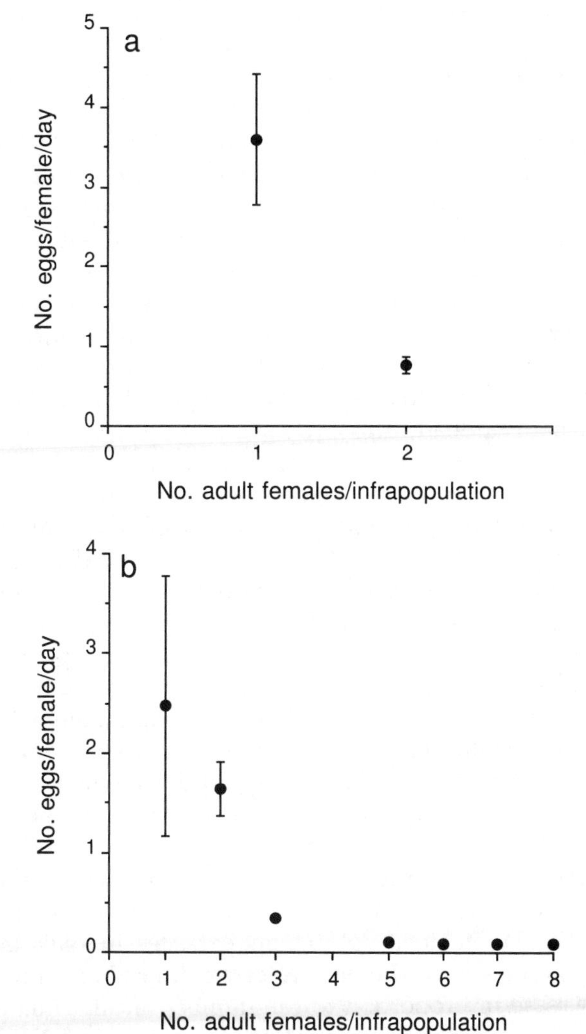

Figure 7.4 Density-dependent egg production by adult female nematodes in insects. Egg output (mean ± standard deviation) of (a) *Blatticola monandros* in the cockroach *Parellipsidion pachycercum*, and (b) *Protrellus dixoni* in the cockroach *Drymaplaneta variegata*, as a function of infrapopulation size. There is typically only one male per infrapopulation in both systems. (Data from Zervos, 1988a, b.)

worms are successful at establishing in the first exposure than in the second. Unlike the acanthocephalan example presented earlier, this is not a case of competition for a limiting resource. Instead, the first infection triggers the production of host antibodies that do not affect the recently established worms but that will attack any incoming cercariae (Smithers and Terry, 1969; Terry, 1994). Similarly, digenean cercariae can face density-dependent host responses when trying to encyst in their second intermediate host. Cercariae of *Cryptocotyle lingua,* for instance, are capable of encysting in cod until the fish host begins to mount an immune response against new cercariae when infrapopulations reach a threshold size (Lysne *et al.*, 1997). The effect of concomitant immunity or other forms of host restriction on the parasite population may be indistinguishable from actual competitive interaction between parasites.

Parasite-induced host mortality can also serve to regulate parasite populations in a density-dependent manner, but it differs from the other mechanisms in at least two ways. Firstly, instead of causing a partial reduction in the number of individuals in an infrapopulation or in their average egg production, it leads to the death of the entire infrapopulation and its loss from the overall population. Because larger infrapopulations are the most likely to disappear (if the probability of host death is a function of parasite load), this process can have major impacts on the parasite population. Secondly, mortality removes individual hosts from the host population, and thus affects host population dynamics as well. Despite the numerous laboratory examples of large infrapopulations killing their hosts, there are few field demonstrations that hosts harbouring more parasites are more likely to disappear (Gordon and Rau, 1982; Adjei *et al.*, 1986; Hudson *et al.*, 1992). It is important to note that in many cases host death is necessary for the transmission of the parasite, and that its effect on the parasite population as a whole may be null: the infrapopulation is not lost, it is transferred from one compartment of the population (e.g. juvenile parasites in the intermediate hosts) to another (adults in the definitive host). Parasite-induced host death only serves as a regulating mechanism when it also kills the parasites in the infrapopulation (e.g. Hudson *et al.*, 1992).

As suggested in the preceding paragraph, and despite the nature of the examples used so far, there is no reason to believe that density-dependent regulation does not act on parasites in their intermediate hosts. Many studies have shown that cercarial production in snail hosts does not increase in proportion to the number of miracidia penetrating those snails (Touassem and Théron, 1989; Gérard *et al.*, 1993). This means that either the establishment success of miracidia and/or their subsequent asexual multiplication are regulated in a density-dependent fashion. The much-studied cestode *Hymenolepis diminuta* provides another example. Regulation of the adult parasite in its definitive rodent host is well documented, and consists of density-dependent effects on worm establishment, survival, growth and fecundity (Hesselberg and Andreassen, 1975; Keymer *et al.*, 1983). In the beetle intermediate host, however, things are different (Figure 7.5). Establishment success is not dependent on density, even at infrapopulation sizes which normally result in the death of the intermediate host (Keymer, 1981). Growth of the larval stages in the intermediate host, on the other hand, is clearly density-dependent. As body size may determine how likely larvae are to become established in the definitive host and their future reproductive success, this may be an instance of density-dependent population regulation operating among larval stages in the intermediate hosts. Similar density-dependent effects in the intermediate host have been reported in other cestode species (e.g. Rosen and Dick, 1983; Nie and Kennedy, 1993). In fact, density-dependent regulation could also act on the free-living infective stages in the external environment, if they are subject to predation by non-hosts.

Most of the examples of density dependence currently available come from laboratory studies. It is possible that in nature, infrapopulation sizes are never large enough for intraspecific competition to occur, and that threshold densities are achieved only in experimental infrapopulations. This would be consistent with the non-equilibrium view of parasite populations espoused by Price (1980). The largest infrapopulations observed in nature could either be what is left of larger infrapopulations following the action of density-dependent regulation, or they could be the unusual occurrence of many conspecific parasites in the same host, which nonetheless remains below the threshold density for the onset of density dependence. The presence of interference competition mechanisms in many species indicates that intraspecific competition has played a role in the evolution of some parasites, and that it occurs in nature as well as in the laboratory.

Density dependence does not occur in all parasite populations, however (see examples given by Combes, 1995). Still, the few examples presented above and others (see reviews by Keymer, 1982; Quinnell *et al.*, 1990; Shostak and Scott, 1993) suggest that density-dependence is common among host-parasite systems. There have been suggestions, however, that many apparent cases of density dependence may in fact be statistical artefacts (Keymer and Slater, 1987; Shostak and Scott, 1993). For instance, the variability in the *per capita* egg production of parasites in small infrapopulations is often higher than that of parasites in larger infrapopulations (see Figure 7.4). A depression in individual egg output in large infrapopulations could therefore result from the chance occurrence of many parasites with low inherent reproductive capacity. A re-examination of published results, using a simulation technique that takes into account the variability among individual parasites (Shostak and Scott, 1993), generally supports the existence of density-dependent

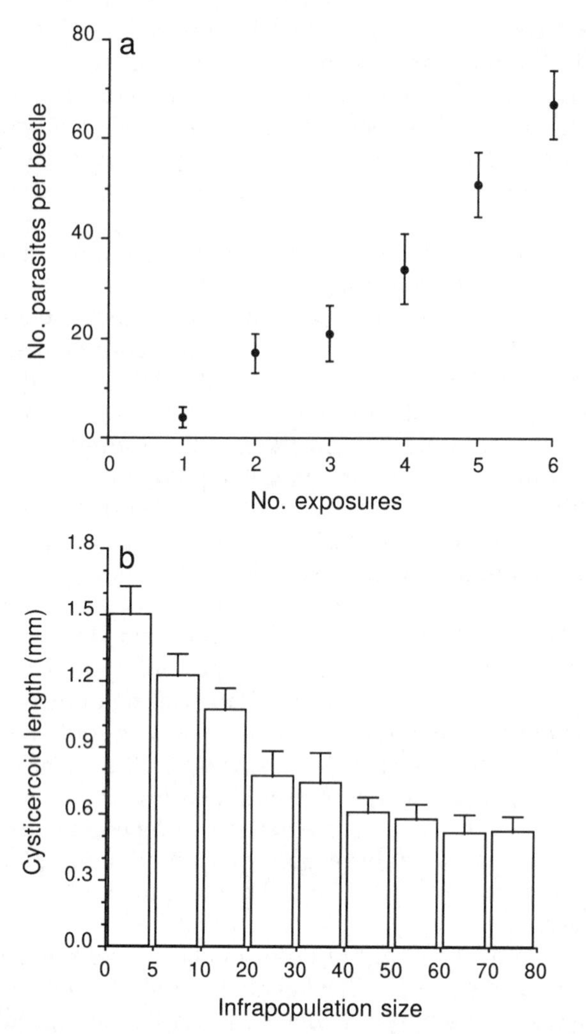

Figure 7.5 Density-dependent effects on establishment and growth of cysticercoids of the cestode *Hymenolepis diminuta* in experimental infections of its intermediate host, the beetle *Tribolium confusum*. (a) Number of established parasites (mean ± 95% confidence limits) after beetle hosts had been put through different numbers of exposures to 1500 parasite eggs; the almost linear relationship suggests that establishment is not dependent on density. (b) Length (mean ± 95% confidence limits) of cysticercoids as a function of infrapopulation size, clearly showing that growth is density-dependent. (Modified from Keymer, 1981.)

mechanisms acting in infrapopulations. This form of regulation is probably operating in many parasite populations and contributing to their long-term stability.

7.3 SELECTED EXAMPLES OF POPULATION STUDIES

Density-dependent processes acting within infrapopulations may be capable of regulating parasite populations, but demonstrations in natural populations are difficult. The myriad environmental factors also acting on parasite numbers make the presence of regulation difficult to detect, or to discern from the effects of abiotic factors operating independently of density. The following examples illustrate how parasite population abundance results from the complex interactions of both density-dependent and density-independent factors in diverse fish-parasite associations.

7.3.1 THE CESTODE *BOTHRIOCEPHALUS ACHEILOGNATHI*

The cestode *Bothriocephalus acheilognathi* has a two-host life cycle involving a copepod intermediate host and a fish definitive host. Infection of fish occurs via predation on infected copepods. The cestode exploits a broad range of fish species, but does not achieve the same reproductive success in all host species. A long-term study of a population of this cestode in a North Carolina lake clearly demonstrates how external factors interact with density-dependent processes to determine population abundance. Typically, prevalence and mean infrapopulation size are lowest in summer months, rise sharply in autumn, peak in winter and decrease in the spring (Granath and Esch, 1983a). These changes are controlled in part by seasonal events in the life cycle of the parasite. Recruitment of new worms by fish takes place only from late spring to the autumn, when copepods are abundant and acquire the infective stages hatched from eggs. Also, during early summer, many adult worms have reached the end of their life and are lost from infrapopulations.

These events are not enough, however, to explain the decline in infrapopulation sizes occurring in spring and early summer. Granath and Esch (1983b) suggested that the rise of temperature in the spring triggers the growth and maturation of cestodes and that this is accompanied by an intensification of exploitation competition for space or nutrients. In winter, almost all parasites are immature, unsegmented worms only a few millimetres long. As they grow rapidly in the spring toward an adult size of a few centimetres, many are eliminated by competition in the densest infrapopulations. Therefore, the population is regulated by density-dependent survival through exploitation competition, which in turn is initiated by a density-independent, seasonal change in ambient temperature.

The study population allows for the role of density-independent, abiotic factors to be examined further. The lake in which the study was performed is used as a cooling reservoir for a power station, and parts of it are thermally altered. In these areas, water temperature fluctuates seasonally but within a higher range, with summer values reaching almost 40°C. Recruitment of new parasites begins sooner in spring and lasts longer into the autumn in thermally altered areas of the lake, but is interrupted for several weeks in summer when water temperature exceeds 35°C because fish predation on copepods declines. As a result, average infrapopulation sizes were smaller in the thermally altered areas than in the unaffected areas of the lake. In addition, effluents from the power plant created a gradient in selenium pollution within the lake. Adult worms in selenium-contaminated hosts produced fewer and less viable eggs than worms in hosts from unpolluted sites (Riggs and Esch, 1987; Riggs *et al.*, 1987). These results clearly demonstrate that external factors can override density-dependent processes by preventing infrapopulations from reaching sizes at which competition becomes important.

7.3.2 THE NEMATODE *CYSTIDICOLA CRISTIVOMERI*

The nematode *Cystidicola cristivomeri* also has a two-host life cycle. It uses a single species of mysid shrimp, *Mysis relicta*, as intermediate host. The life cycle is completed after an infected intermediate host is ingested by a suitable fish host (*Salvelinus alpinus* or *S. namaycush*). There the nematode migrates to the swim bladder of the host where it matures (Black and Lankester, 1980). This parasite is long-lived, and may survive more than 10 years in the definitive host. For this reason, the dynamics of infrapopulations do not show seasonal fluctuations but must be examined over longer time spans.

Typically, each infected shrimp harbours a single infective larva. Therefore, a single new recruit is added to an infrapopulation in a fish each time the host consumes an infected prey. Mysid shrimps are important food items in the diets of small host fish, but tend to become secondary as fish grow and become increasingly piscivorous. As a consequence, recruitment rates are high in young infrapopulations harboured by young fish hosts, but decrease with age. The pattern observed in natural populations shows an increase in mean infrapopulation size with increasing fish age up to a certain age, which varies among populations in different lakes, after which infrapopulation size stabilizes (Black and Lankester, 1981). This probably results from a parasite recruitment rate higher than the parasite mortality rate in the first few years of an infrapopulation's existence, followed by a period in which both rates are roughly identical until the death of the host.

Whereas infrapopulations may not often reach sizes at which parasite survivorship becomes negatively affected, there is evidence of density-dependent reproductive output in this nematode. Both the proportion of female nematodes reaching sexual maturity, and the average length of females, are inversely related to infrapopulation size.

As fecundity is strongly correlated with body length in *Cystidicola* nematodes (Black and Lankester, 1981; Black, 1985), *per capita* egg production is reduced in large infrapopulations. Thus both truly density-dependent effects and temporal changes in recruitment rates, which coincide with increases in infrapopulation size but are dependent only on host feeding behaviour, contribute to the long-term stability of the nematode population.

7.3.3 THE NEMATODE *CYSTIDICOLOIDES TENUISSIMA*

The nematode *Cystidicoloides tenuissima* is a common parasite of salmonid fish in the Holarctic. Adults are found in the stomach of their fish definitive hosts and juveniles live in the body cavity of aquatic insects, particularly mayflies, from where they reach the definitive host when the latter ingests infected intermediate hosts.

In the River Swincombe in England, where adult parasites infect brown trout, *Salmo trutta*, and juvenile Atlantic salmon, *S. salar*, maturation of the parasite is strongly linked with water temperature. Gravid female worms are present, and eggs are released in the water, from summer to early winter (Aho and Kennedy, 1984). The timing of egg release coincides with the appearance in the river of a new cohort of the mayfly *Leptophlebia marginata*, the only species in the River Swincombe in which the nematode can develop to become infective to fish. Insect hosts become infected by eating nematode eggs throughout the summer and autumn; the majority of eggs ingested by insects are not eaten by *L. marginata* (Figure 7.6). Nematode juveniles acquired by mayflies in the summer soon become infective to fish; those acquired later may reach infectivity only in the spring, as there is little or no development during the winter months.

Trout consume more mayflies than salmon, and as a consequence they harbour most of the adult parasite population (Figure 7.6). Fish hosts do not feed much on *L. marginata* during the summer; the mayfly becomes more common in fish diet in autumn and throughout winter and spring. Consequently, the mean infrapopulation size of adult parasites in fish increases beginning in the winter and peaking in spring before declining rapidly in summer. At that time, not only does recruitment stop until the autumn, but also adult parasites start dying. There is no evidence for density-dependent competition acting in this system. The onset of high mortality of adults coincides with a peak in infrapopulation sizes, but appears to be caused by a temperature-dependent host response (Aho and Kennedy, 1984). Thus, in this system, infrapopulation sizes are controlled by a density-independent rejection induced by high summer temperatures, and are kept below the threshold, if any, at which density-dependent mechanisms would start operating.

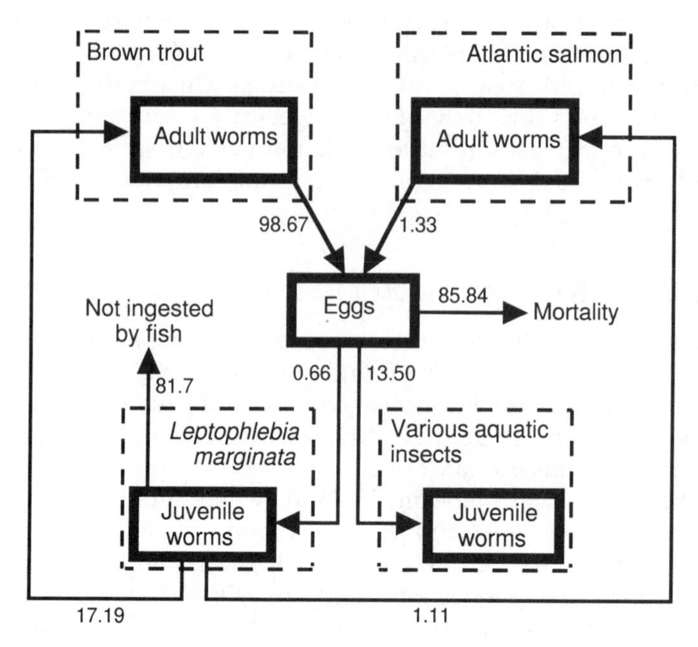

Figure 7.6 Summary of the circulation and transmission dynamics of the nematode *Cystidicoloides tenuissima* in the River Swincombe, England. All numbers are percentages. Most adult worms live in trout, and thus trout parasites release many more eggs than salmon parasites. The majority of eggs die or are ingested by insects in which they cannot develop further. Of the very few that are eaten by the mayfly *Leptophlebia marginata*, the only suitable intermediate host, only 18.3% ever establish in definitive hosts, with the majority being recruited by adult infrapopulations in trout. Numbers are from one site in the river but are representative of other sites. (Data from Aho and Kennedy, 1987.)

7.4 SUMMARY

The population biology of parasites has been investigated on two fronts. Firstly, the dynamics of parasite populations can be modelled mathematically, with a few simplifying assumptions. This is the epidemiologist's approach, and it has proven to have predictive value. Secondly, empirical studies of field populations have highlighted the many density-dependent and density-independent mechanisms acting to regulate parasite abundance over time in specific systems. This is the approach of the ecologist, and it has served to validate some aspects of models.

But from the evolutionary biologist's perspective, some of the main questions remain completely unanswered, even unasked. For instance, what is the evolutionary significance of density dependence? Is it an unavoidable consequence of crowding, or have the mechanisms behind it been actively shaped by selection? Clearly, in cases of interference competition, selection has favoured individuals capable of eliminating their competitors. But even in situations where exploitation competition is the operating mechanism, why is density dependence detected in some species but not in others? When it is present, why does it manifest itself at different densities among closely related species? Is the occurrence of density dependence related to parasite phylogeny? It may be that density-dependent mechanisms have evolved differently in various parasite lineages. There are virtually no comparative analyses of population-level phenomena among parasite species. Ecologists studying free-living organisms have performed several such analyses but have not often considered possible phylogenetic effects (Harvey, 1996). As more examples of density-dependent regulation in parasite populations become available, we may be able to tackle these important questions.

8 Interactions between parasite species

Previous chapters have addressed several aspects of parasite ecology as though parasites of one species were alone in their host. This is rarely the case as most free-living animals, especially vertebrates, are used as host simultaneously by several species of parasites. Pairs of these different parasite species, though sharing a host, will not always co-occur on a finer scale: an intestinal worm and an ectoparasitic arthropod on the same vertebrate host will not interact and can be treated as entirely independent of one another. Often, however, hosts harbour several parasite species belonging to the same guild, i.e. parasite species that use similar resources such as food or space. These species are likely to interact with one another regardless of differences in other aspects of their biology. This chapter examines interactions among parasite species, and serves as a bridge to the next two chapters which deal with parasite communities.

Several types of interspecific interactions are possible among parasites within the host. The effect of one species on another can be positive. For instance, interference with host defence mechanisms by one parasite species can facilitate the exploitation of the host by a second species. More often, however, interactions will be antagonistic such that the presence of one parasite species has negative effects on the numbers, distribution or reproduction of other species. These antagonistic interactions can include predation; for example, the larval stages of digeneans in snails can prey on smaller larval stages of other digenean species (Sousa, 1992, 1993). The most commonly reported interactions and the most studied, however, consist of various forms of interspecific competition.

The sort of evidence used to infer that competition for resources occurs between two species is varied (Thomson, 1980). Firstly, a reduction in the number of individuals of one species when the other is present suggests that competition is taking place. Secondly, individuals of one species can change the way in which they use a resource when the second species is present. This is possible because of some plasticity in morphology or behaviour; among parasites, this would often take the form of a slight shift in the site of infection. Both these phenomena, the numerical and functional responses to the presence of a competitor, are indications that two species affect one another. Ecologists often feel that numerical responses are more convincing demonstrations of competitive interactions than functional responses (Thomson, 1980). Here, equal weight is given to both types of evidence, and they are discussed in turn. Cases are also considered in which similar parasite species do not interact because of small but consistent differences in resource use. These may represent evolutionary niche shifts or character displacements resulting from intense competition in the past; they may however be the result of other evolutionary forces and the fact that they prevent competition may be a fortuitous consequence.

8.1 NUMERICAL RESPONSES TO COMPETITION

As emphasized in the previous chapter, the regulation of parasite populations occurs through mechanisms acting in infrapopulations. Similarly, interactions between parasite species take place within individual hosts where two or more species co-occur. To demonstrate that two parasite species compete, it is necessary to show a negative relationship between the numbers of individuals of species A and species B among a sample of host individuals. This can be done using a sample of naturally infected hosts, or ideally using experimental infections. The experimental approach allows the infrapopulation sizes of one species alone in the host to be compared with its infrapopulation sizes when it shares the host with a presumed competitor, while doses of infective larvae and other variables are kept constant.

Dobson (1985) reviewed the literature on concurrent infections of parasites, and found that reductions of infrapopulation sizes by as much as 50% were common in mixed infections of gastrointestinal helminths in mammals. Examples of numerical effects of competition among closely related and unrelated helminths of mammals are shown in Figure 8.1. Typically, competition between species is asymmetrical, with one species suffering severe losses and the other being almost unaffected. In other words, the infrapopulation of the winning species is unchanged but it reduces the size of the infrapopulation of the losing species, sometimes excluding it entirely from the host individual. This indicates one-sided interference rather than exploitative competition (as defined in Chapter 7), but can also result from host-mediated effects involving immune responses. Similar asymmetrical interactions are also observed in mixed infections of intermediate hosts. For instance, the establishment success of cysticercoids of the cestode *Hymenolepis diminuta* in its insect intermediate host is severely reduced by the presence of another cestode,

Raillietina cesticillus, but the reciprocal effect is not observed (Gordon and Whitfield, 1985). Also, in a comprehensive review of published studies of larval digeneans in snail intermediate hosts, Kuris and Lafferty (1994) showed that subordinate digenean species suffered severe losses from

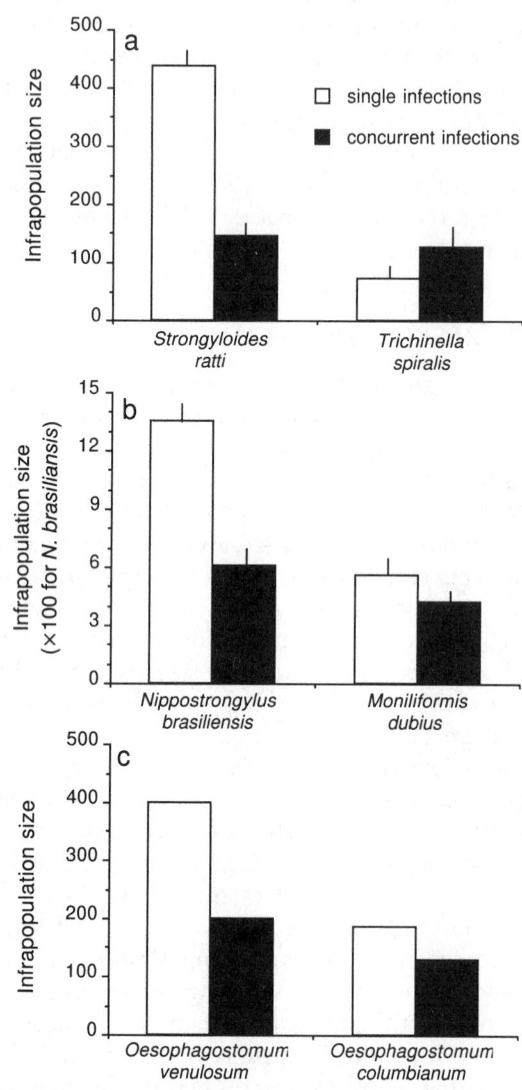

Figure 8.1 Infrapopulation sizes (means + standard errors, if available) of helminths of mammals in single and concurrent infections in experimental studies. (a) Two nematode species in rats; (b) a nematode and an acanthocephalan in rats; (c) two congeneric nematode species in sheep. The outcome of competition is typically asymmetrical, with only one species (on the left) suffering a significant numerical decrease. (Data from Moqbel and Wakelin, 1979; Dash, 1981; Holland, 1984.)

interspecific antagonism compared to dominant species (Figure 8.2). The main determinant of a species rank within these hierarchical interactions appears to be size, with the larvae (rediae) of large sizes preying on and excluding larval stages of other digenean species from their snail host (Sousa, 1992, 1993).

Numerical responses other than reductions in infrapopulation size are also possible. Often the species suffering the most from interspecific competition will incur a reduction in fecundity when co-occurring in a host with its competitor (e.g. Moqbel and Wakelin, 1979; Silver *et al.*, 1980; Holland, 1984). These can lead to a decrease in recruitment rates and affect the parasite population as a whole. As with effects on survival, effects of competition on fecundity are often asymmetrical, with one species affected much more than the other. It must be pointed out, however, that the outcome of competition may be influenced by the order in which parasite species become established in the host. The species that suffers most from competition when it enters the host simultaneously with a competitor can perform much better if given a head start, i.e. if

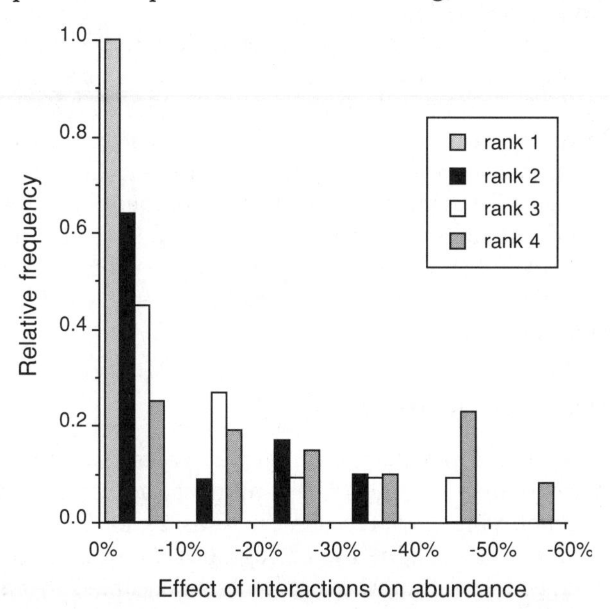

Figure 8.2 Relative effect of antagonistic interactions on the abundance of individual parasites of different digenean species exploiting the same snail population. The data come from 25 studies on littorine snails. The relative (percentage) change in abundance compared to infrapopulations not sharing their snail host with other species was calculated for each species in each study. Species were ranked in order of competitive dominance (rank 1 being most dominant) based on evidence available for each host-parasite system. Dominant species incur very small reductions in numbers, if any, due to interspecific interactions; subordinate species regularly suffer severe losses. (Modified from Kuris and Lafferty, 1994.)

allowed to establish in the host well before the competing species arrives. Priority effects have been observed in most of the systems where asymmetrical competition occurs in concurrent experimental infections.

Despite the often large numerical effect of interactions among parasite species in experimental infections, the magnitude of interspecific competition between parasites in nature may often not be important. In the laboratory, large infrapopulations of different species are placed in the same host individuals, a situation that does not arise often in natural systems. The reason for this is that the typical distribution of parasites among hosts creates few opportunities for two or more species of parasites to co-occur in the same host individual in sufficient numbers for competition to become measurable. Assuming that different parasite species display aggregated distributions, and that these distributions are independent of one another, potentially severe antagonistic effects may never be realized and parasite populations may coexist in the same host population. The restraining influence of aggregation on interspecific competition is also predicted by mathematical models (Dobson, 1985; Dobson and Roberts, 1994).

When moderate numbers of parasites of different species end up in the same host, they can also alter their resource use and avoid the numerical effect of competition; this will be addressed in the following sections. In recent years parasitologists have devoted more effort to the study of functional responses to competition than to the measurement of numerical effects. Inferring the existence of competition from functional responses without demonstrating numerical effects can be misleading (Thomson, 1980), but the many examples of niche adjustments shown by co-occurring parasite species (see section 8.3) suggest that these are indeed responses to antagonistic interactions.

8.2 THE PARASITE NICHE

If competition results in a change in the way parasites use host resources more often than in a change in the number of parasites, we need a method of quantifying resource use by parasites to detect competition. One approach is to determine the ecological niche of individual parasite species, which is the multidimensional volume occupied by parasites and defined by several physical and biotic variables (Hutchinson, 1957). Most of these variables, such as the exact type of food particles ingested by parasites, are extremely difficult to quantify, just as they are for the majority of free-living animals. Parasitologists have therefore focused on the spatial dimension of the niche. Just as they tend to be highly host specific (see Chapter 3), parasites are restricted to particular sites on or in the host. Detailed measurements of parasite attachment sites can be made and used to compare the niches of parasites in single and mixed infections.

These measurements can be simple to obtain. If the parasite habitat can be defined along a linear axis, such as the vertebrate gastrointestinal tract extending from the oesophagus to the anus, then a measure of the niche can be taken as the mean or median position of individual parasites. The niche itself would be the region encompassed by the range of their positions. Despite many simplifying assumptions (Bush and Holmes, 1986b), this one-dimensional approach provides useful information on parasite niches. Indices of niche width and other n-dimensional measures of site specificity have also been used (e.g. Anderson *et al.*, 1993; Rohde, 1994b). These or other more complicated measures may be necessary when parasites occupy structurally more complex microhabitats in the host.

Defining the spatial boundaries of a parasite's distribution in its host leads one to believe that there must be a maximum number of parasite species, or parasite niches, that can be fitted in a host. How can we estimate the number of niches in, say, an intestine? Dividing the length of the intestine by the average niche width of parasites does not provide an answer as there are gradients in habitat quality along the intestine and some parts may be unsuitable for all parasites. Perhaps the simplest way is to take the maximum number of species co-occurring in one host as the number of niches available in hosts of that species (e.g. Kennedy and Guégan, 1996). This can only be a lower estimate, however, as it assumes that hosts are saturated with parasite species and that there are no niches left unoccupied.

A distinction must be made between the fundamental niche and the realized niche. Strictly in terms of spatial location in the host, the fundamental niche is defined as the potential distribution of the parasite in the host, i.e. the range of sites in which the parasites of one species can develop. Realized niches are subsets of the fundamental niche, and consist of the sites actually occupied by parasites in the host. Realized niches represent either the optimal sites within the fundamental niche if interactions with other species are unimportant, or the portions of the fundamental niche actually available to them because of antagonistic interactions with other parasite species. The best way to measure the fundamental niche of parasites would be to quantify their distribution in controlled, single-species infections (e.g. Holmes, 1961; Patrick, 1991). As this is often impossible, another way to estimate the boundaries of the fundamental niche consists of summing up the distributions of parasite individuals across all hosts in which the species occurs, in a sample of naturally infected hosts containing mixed infections (Bush and Holmes, 1986b). The functional response of parasites to interspecific competition can then be examined by comparing the realized spatial niche of infrapopulations alone in the host with that of infrapopulations sharing the host with presumed competitors.

8.3 FUNCTIONAL RESPONSES TO COMPETITION

Holmes (1973) refers to adjustments in infection site in response to the presence of a competitor as interactive site segregation. This happens when the fundamental niches of two parasite species overlap: one or both of them adopt sites other than their preferred sites, but still within their fundamental niche, in order to minimize the overlap and thus competition. This requires that parasites are able to detect heterospecifics or at least a change in the environment caused by heterospecifics, and that their behaviour is plastic enough to adjust by changing the site of attachment. A slight shift away from the preferred site can be associated with a decrease in fitness if conditions in the alternative site are sub-optimal. The cost of competition must therefore outweigh the cost of this shift in attachment site for interactive site segregation to be favoured by selection. Indeed, substantial spatial niche overlap does not mean that competition occurs, even between related species (e.g. acanthocephalans in eels; Kennedy and Moriarty, 1987; Kennedy, 1990), and site segregation will not be favoured if competition is not costly. If competition is very severe, however, if it is maintained over several generations, and if competing species are very likely to co-occur in the same host within each generation, we may expect the niche shift to become genetically fixed. This would mean a reduction of the overlap between fundamental niches instead of between realized niches, and has been labelled by Holmes (1973) as selective site segregation. This evolutionary phenomenon will be considered further in the following section; here the discussion centres on proximate adjustments to the realized niche in response to the immediate presence of a competitor.

Intestinal helminth communities in many birds often consist of large numbers of individuals of several species, and provide good evidence for functional responses to the presence of competitors. In four species of grebes, for instance, Stock and Holmes (1988) found that each helminth species was usually restricted to a predictable portion of the intestine and that the various species were arranged along the entire length of the intestine. There was much overlap between the fundamental spatial niches of the different parasite species, however, when the fundamental niche of each species was determined from the pooled distributions of conspecific parasites in all hosts (Figure 8.3). A reduced overlap between the realized spatial niches of pairs of helminths co-occurring in individual birds relative to that between their fundamental niches would be evidence for interactive site segregation as a functional response to competitors. Stock and Holmes (1988) indeed found that among all pairs of species with fundamental niches that overlap by at least 5%, significant reductions in the overlap of realized niches were the general rule.

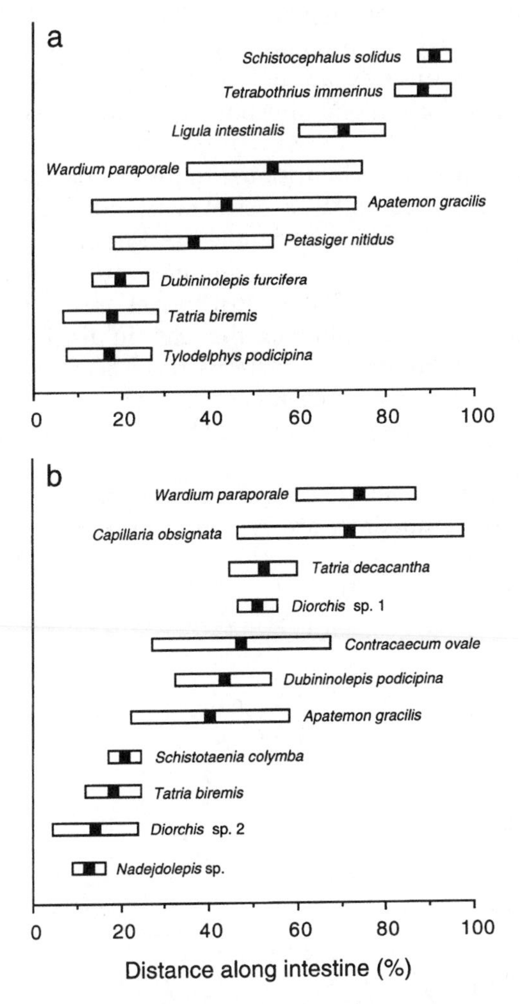

Figure 8.3 Fundamental niches of intestinal helminths in two species of grebes: (a) *Aechmophorus occidentalis* and (b) *Podiceps nigricollis*. The mean position (± standard deviation) along the length of the intestine is the average of the positions of the medianth individual in each infected host individual. Helminth species with prevalences smaller than 25% are not shown. (Data from Stock and Holmes, 1988.)

Bush and Holmes (1986a, b) also found evidence of functional responses to interspecific interactions in rich communities of helminths in birds. Among the more common species of intestinal parasites of the lesser scaup duck, *Aythya affinis*, overlaps of realized spatial niches between pairs of species were typically less than half those between fundamental niches (Figure 8.4). Interestingly, whereas the breadth of the realized niche of many species within individual hosts increased with infrapopulation size, the overlap between the realized niches of adjacent species did not increase with their combined infrapopulation sizes. This was due to asymmetrical expansions of the realized distribution in the intestine, i.e. expansions in one direction only, away from the main competitor's niche.

Spatial niche adjustments in response to competition occur in parasites other than intestinal helminths of vertebrates, although the latter have received more attention in past studies. Arthropods ectoparasitic on birds, for instance, often form communities rich in individuals and species (Choe and Kim, 1988, 1989). Most species are site-specific, and the presence of one species can influence the realized distribution of another species (Figure 8.5). Similar observations have been made on oxyurid nematode species parasitizing cockroaches (Adamson and Noble, 1992), and interactive niche segregation no doubt occurs in other types of parasite community.

Many more examples could be given of an apparent narrowing of realized niches when competitors are present and of reductions in spatial

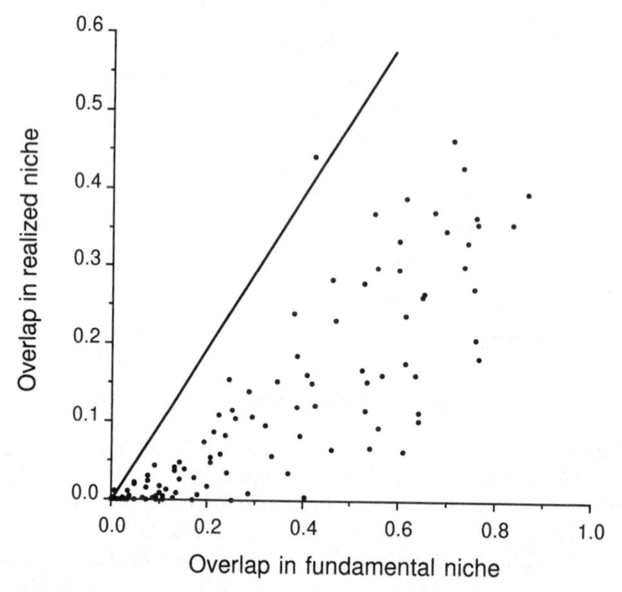

Figure 8.4 Relationship between the overlap in realised niche and the overlap in fundamental niche in all pairwise associations (120 in total) among 16 intestinal helminth species in the lesser scaup duck, *Aythya affinis*. All parasite species were common ones, found in at least one-third of birds examined. The niche is defined as the linear distribution of the parasites in the intestine, and overlap between the niches of two species is measured as the proportion of the individuals of the two species found in the area of overlap. Points falling below the line represent overlaps between realized niches smaller than between fundamental niches. (Data from Bush and Holmes, 1986b.)

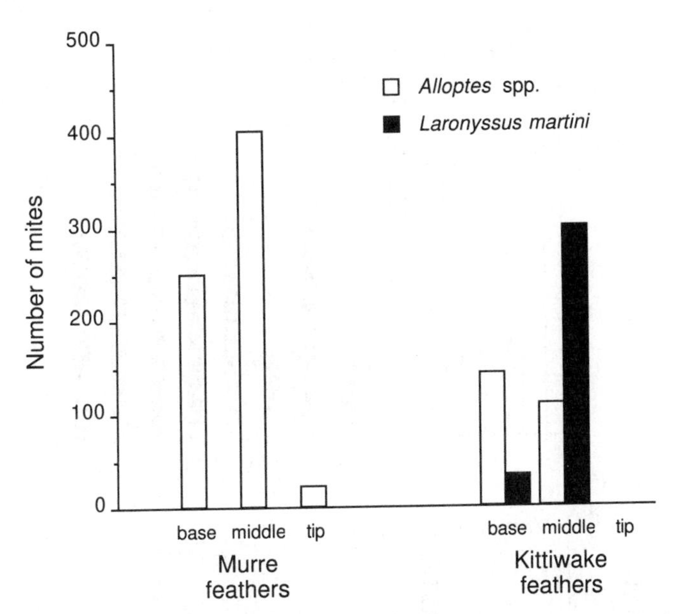

Figure 8.5 Distributions of mites along the length of feathers on murres (*Uria* spp.) and kittiwakes (*Rissa* spp.). A single species of mite of the genus *Alloptes* occurs on murres, where it is found mainly on the middle sections of feathers followed by the base of feathers. On kittiwakes, *Alloptes* occurs with another species, *Laronyssus martini*, which shows a strong preference for the middle sections of feathers. The realized niche of *Alloptes* on kittiwakes differs from that on murres, possibly because of a shift of its sites of attachment toward the base of feathers in response to interspecific competition. (Modified from Choe and Kim, 1989.)

niche overlap (see Combes, 1995). These all suggest that competition is avoided through active site segregation. There exists at least one other explanation for these results when they are obtained from naturally infected hosts. Parasites may select attachment sites along the whole breadth of their fundamental niche, but survive only in those portions of the fundamental niche not occupied by competing species. This would create an illusion of site segregation but would in fact be the simple result of differential mortality along the spatial dimension of the niche. Experimental studies in which known numbers of parasites of different species are used for single and multiple infections of hosts can only focus on a few pairs of species, but provide additional arguments in favour of interactive site segregation (Figure 8.6).

Spatial segregation of the realized niches of competing species may be achieved at a larger scale: a parasite species could reduce niche overlap with competitors by excluding them from the individual hosts in which it has established. This phenomenon is common among larval trematodes exploiting snail hosts (Kuris and Lafferty, 1994). But is this interactive

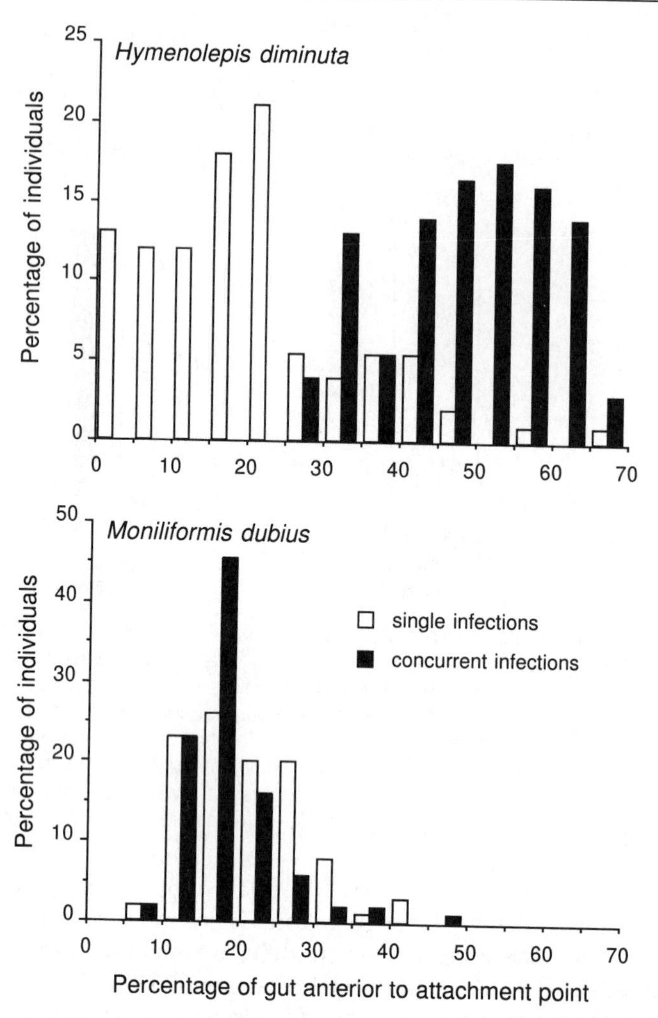

Figure 8.6 Distributions of the cestode *Hymenolepis diminuta* and the acanthocephalan *Moniliformis dubius* along the intestine of rats in experimental infections. The fundamental niches of the two worms, i.e. their distributions in single infections, overlap completely. Their realized niches in concurrent infections, however, show only minimal overlap due to the distribution of *H. diminuta* shifting down the intestine. This is a true example of interactive site segregation and not an artefact of differential mortality along the length of the intestine, as similar numbers of worms of both species were recovered in single and concurrent infections. (Modified from Holmes, 1961.)

exclusion a functional response of parasites to competition, or an extreme numerical effect? The latter is probably the process responsible for the exclusion of one species by another. Parasites are recruited by hosts harbouring competing species and by hosts not harbouring competitors, and they fail to establish and survive in hosts harbouring competitors. Thus partial or total exclusion is a case of interspecific competition but not really one of interactive spatial segregation.

Functional responses along niche dimensions other than spatial location in the host have also been reported. For instance, congeneric species of digeneans of similar sizes have been reported to diverge in body size when occurring together, perhaps in order to minimize the overlap in resource use (Figure 8.7). Such examples reinforce the case that interspecific interactions can shape the spatial or size structure of helminth communities, even if actual competition is avoided through niche shifts.

The importance of interspecific interactions in the above examples is a consequence of the broad fundamental niches of the co-occurring species. This is not a characteristic of all parasite communities. In some communities, called isolationist parasite communities, parasite species have narrow fundamental niches that are often identical to realized niches. In these communities the breadth and position of the niche of one species is independent of the presence of other species (Rohde, 1979; Price, 1980). Interactions play no detectable roles, and the characteristics of the niche have other determinants. Holmes and Price (1986) emphasized that interactive and isolationist communities are at two ends of a

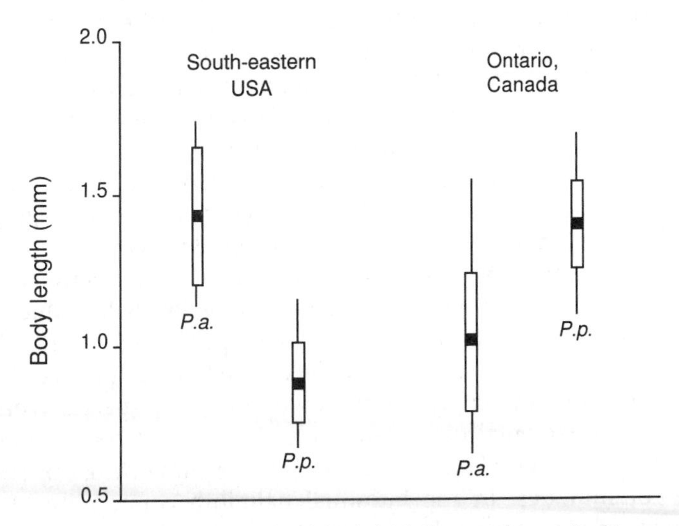

Figure 8.7 Body length (mean ± standard deviation, and range) of two congeneric digenean species parasitic in raccoons, *Pharyngostomoides adenocephala* (*P.a.*) and *P. procyonis* (*P.p.*), in two areas where they occur in sympatry. In areas of North America where they do not co-occur, both species attain only small sizes. When they co-occur, one of them (*P. adenocephala* in the south-eastern USA, *P. procyonis* in Ontario) reaches a significantly larger size than the other. (Modified from Butterworth and Holmes, 1984.)

continuum of parasite communities, such that interspecific interactions range from important to insignificant in determining the sizes and distributions of parasite infrapopulations. The existence of the continuum is open to debate, however, as the detection of species interactions depends entirely on how one chooses to measure interactive effects, i.e. numerical or biomass effects, functional responses, etc. (Moore and Simberloff, 1990). Species interactions, nonetheless, while playing no current roles in at least some communities, may have been important earlier in the evolution of these communities and may have helped shape the modern niche of their component species, as explored in the following section.

8.4 EVOLUTIONARY NICHE RESTRICTION

In some communities, parasite species have restricted fundamental niches which overlap very little or not at all with neighbouring species, and interspecific interactions are unimportant. These narrow and highly specific niches may have had at least two evolutionary origins: they may be the product of intense competition between parasite species in the past; or they may have other origins and have prevented competition from occurring. In other words, either specialized niches are the consequence of intense interspecific competition in the past, or interspecific competition in the past never occurred because of the lack of overlap between specialized niches. Distinguishing between the two scenarios is practically impossible (although comparative studies would probably offer useful insights). Whatever the evolutionary reason for the lack of niche overlap in isolationist communities, however, the influence of present competition in these communities is negligible.

Evolutionary niche restriction driven by interspecific competition can take place in some circumstances (Holmes, 1973). Selective site segregation over evolutionary time will be favoured when abundant and prevalent parasite species share the same host population for numerous generations, and when competition among these species leads to reductions in fitness. If these conditions are met, parasite species will exert selective pressures on one another and will coevolve such that their niches diverge. After enough time, the community would consist of species with non-overlapping fundamental niches and only the ghost of competition past would remain (Connell, 1980). There is evidence supporting this scenario in some communities. In the intestinal helminth community of lesser scaup ducks, *Aythya affinis*, the prevalent and abundant species (or core species; Hanski, 1982) have non-overlapping niches more predictable and more evenly dispersed along the intestine than expected by chance (Bush and Holmes, 1986b). Core species are those most likely to encounter one another and interact if their niches overlap. In contrast, less prevalent and abundant species (satellite species; Hanski, 1982) have niches randomly distributed along the intestine. In other helminth communities in some avian hosts, species most likely to compete because of similar resource requirements, i.e. congeneric helminth species, typically have disjunct niches or niches with minimal overlap (Stock and Holmes, 1988). As this applies to both fundamental and realized niches, an evolutionary niche divergence is the likely explanation.

There are, however, many parasite communities in which interactions are not detectable now, and may never have been important because the various species are present at low abundance and do not co-occur regularly (e.g. Lotz and Font, 1985; Goater *et al.*, 1987; Kennedy and Bakke, 1989). These are characterized by low infrapopulation sizes and low recruitment rates. Realized niches are not influenced by the presence of other species, whether or not fundamental niches overlap. In these isolationist communities, the various species are not abundant enough to exert mutual pressures on one another, and any observed niche restriction has evolved independently of interspecific interactions (Price, 1980, 1987).

Based mainly on observations of monogeneans and other ectoparasites of marine fish, Rohde (1979, 1991, 1994b) proposed that niche restriction in parasites often evolves for reasons other than to reduce the effects of interspecific interactions. Several lines of evidence suggest that interspecific competition has no great ecological or evolutionary importance in these communities. Firstly, many apparently suitable niches are vacant and not utilized. Secondly, the common occurrence of congeneric parasite species on the same fish species (Figure 8.8) suggests that interactions are not strong enough to restrict the number of morphologically and ecologically similar species using the same host (Rohde, 1991). Thirdly, the presence of one species, even at high numbers, usually has no numerical or functional effects on other species, i.e. no detectable effects on their numbers or the extent of their realized niches. Other factors, such as the low mean infrapopulation sizes and the highly aggregated distributions of monogeneans, make it unlikely that large infrapopulations of different species will co-occur on the same host (Rohde, 1991; Reversat *et al.*, 1992). If opportunities for intense competition are few or non-existent, there will be no selection for evolutionary site segregation. Encounters between potential competitors are also rare in other types of parasite communities, such as assemblages of larval digeneans in snails (Curtis and Hubbard, 1993), and co-evolved adaptations against competition are unlikely in these communities as well.

Various observations on monogeneans point to another force acting to restrict the niche of parasites. The probability of mating in monogeneans may be low on average because of the typically low infrapopulation sizes shown by these parasites. Rohde (1991) presented evidence that the realized niches of adult parasites are often more restricted than those of juveniles, and that they become even narrower at the time of mating. In

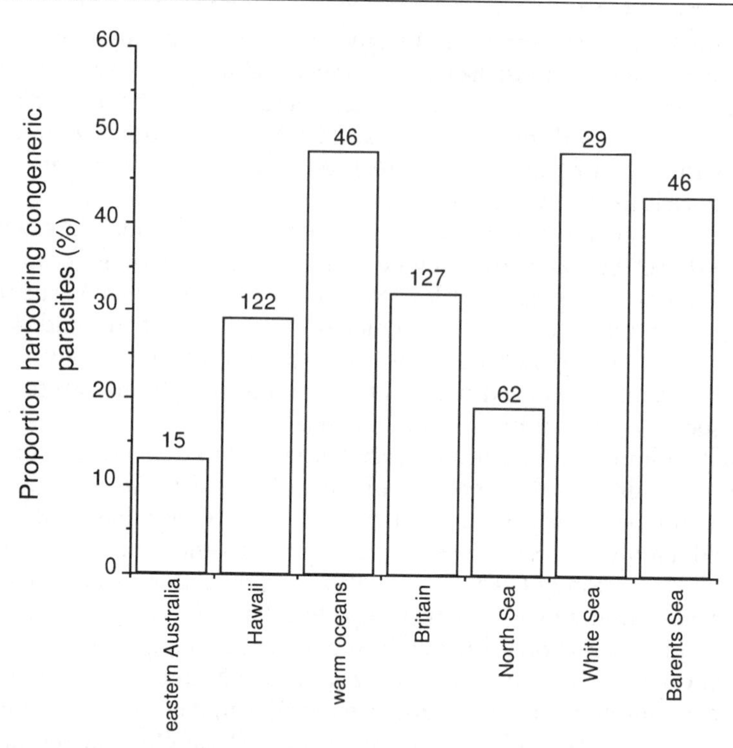

Figure 8.8 Proportion of marine fish species harbouring congeneric species of parasites in seven surveys made in various localities around the world. One survey (Australia) focused on coccidian parasites, the others on metazoans, especially ectoparasites. The number of fish species examined in each survey is indicated above the columns. Results of one survey (not shown) of deepwater fish found no fish harbouring congeneric parasites; however, parasite diversity was very low in that survey. As not all fish species examined were sampled sufficiently, the above values are minimum estimates, but they do suggest that congeners frequently coexist on the same host species. (Data from Rohde, 1991.)

addition, the niches of sessile species are narrower than those of related but more mobile species (Rohde, 1991, 1994b). These and other observations suggest that niche restriction serves to increase the chances of encountering a potential mate in low-density infrapopulations.

Reproduction may therefore have played a role as great as or greater than competition in the evolution of restricted niches in parasites. There are at least two other ways in which the reproductive biology of parasites may have driven the restriction of the niche. Firstly, the overlap between the realized niches of congeneric species of fish ectoparasites is generally lower than that between non-congeners (Rohde, 1991). As all species, whether congeners or not, use the same resources, this observation cannot easily be reconciled with site segregation as a means to avoid competition. Niche restriction may instead serve to maintain or reinforce reproductive barriers between similar parasite species. This conclusion is supported by the interesting observation that congeneric monogenean species with overlapping niches have copulatory organs that differ much more in shape or size than those of congeneric species with segregated distributions on the host (Rohde, 1991, 1994b). If niche restriction evolved to reduce interspecific competition, surely it would have acted similarly on all congeneric species regardless of the differences in their copulatory organs.

The second way in which reproduction may have shaped parasite niches has to do with gradients along the niche dimensions and their effects on the reproductive success of parasites. Many types of parasites achieve greater reproductive success in some portions of their fundamental niche than in others (e.g. Sukhdeo, 1990a; Chilton *et al.*, 1992). It is possible that these differences in fitness have favoured a narrowing of the niche around sites where fitness is maximized.

The general role of interspecific competition in the evolution of restricted parasite niches is probably not negligible, but is difficult to demonstrate. No experiments can separate the effects of competition, reproduction or other factors; the above arguments are all based on observations, mostly of natural populations. Still, competition will exert selective pressures on the niche dimensions of two species sharing the same host species when their abundances and distributions among hosts make them likely to co-occur frequently in large numbers in the same host individual; their fundamental niches overlap significantly; and one or more host resources required by both species is limiting. These conditions are probably satisfied by many communities of intestinal helminths of vertebrates, even species-poor ones if the right circumstances prevail. Studies on niche restriction are still few, and generalizations cannot be made at this point, but it is likely that interspecific competition has played a role in some types of parasite assemblages but not in all of them.

8.5 SUMMARY

Demonstrating the existence of past or present interspecific interactions among parasites is an important step toward understanding the structure of parasite communities. It is a not an easy step, however. Numerical or functional responses observed in experimental infections may be artefacts of infrapopulation sizes larger than natural ones, and their interpretation requires caution. Similarly, the absence of overlap between the fundamental niches of two parasites may be the result of forces other than past competition. These caveats still allow some conclusions to be drawn. When competition does occur, numerical effects are often asymmetrical, with one species bearing the brunt of the competition. Often, but not always, realized niches are narrower than fundamental niches, presumably to reduce overlap with the niches of other species. When niche segregation has evolved to fixation, past competition may only be the cause if opportunities for competition existed among ancestral species.

Assemblages of parasites sharing a host may have a predictable structure if the component species interact with one another, but are likely to be random collections of species if interactions are negligible. It is the interactions between individual parasites of different species in the same host that may determine the species composition of communities and the relative abundance of different species. The next chapter picks up from this one and examines patterns of species associations in various parasite communities.

Parasite infracommunity structure

9

The assemblage consisting of all parasites of different species in the same host individual, whether they interact or not, forms an infracommunity (Holmes and Price, 1986). Infracommunities are subsets of the component community, which consist of all parasite species exploiting the host population. The composition of infracommunities, in terms of the number and identity of species and the relative numbers of individuals of each species, will depend on many factors. In theory, infracommunities can range from highly structured and predictable sets of species to purely stochastic assemblages of species coming together entirely at random. Interactions among parasite species are one of the main forces that can shape an infracommunity and give it a structure departing from randomness. We may thus expect more predictable groupings of species if competitive interactions are strong. In isolationist parasite communities, where interactions are negligible either because of very narrow niches or small infrapopulation sizes, the co-occurrence of species in hosts is not expected to deviate from that expected by chance if interspecific interactions are the main structuring processes in parasite infracommunities.

There is at least one other factor influencing infracommunity structure. Infracommunities are typically short-lived, their maximum life span being equal to that of the host. They are also in constant turnover of parasite individuals, with new ones being recruited and old ones dying out all the time. The probability of each parasite species being recruited into an infracommunity, and the way in which they join infracommunities, will also affect the composition and size of infracommunities.

The aim of this chapter is to review the evidence that the structure of parasite infracommunities differs from that of random assemblages, and to discuss and evaluate the role of interspecific interactions and parasite recruitment patterns in structuring infracommunities. The relative ease with which several host individuals from the same population can be sampled facilitates statistical testing for significant departures from random species assemblages. The lack of adequate null models for comparisons with observed patterns has plagued some earlier studies of parasite infracommunities (Simberloff, 1990; Simberloff and Moore, 1996). This chapter emphasizes the need to contrast observed patterns with those predicted by proper null models.

9.1 SPECIES RICHNESS OF INFRACOMMUNITIES

As each infracommunity is a subset of the parasite species present in the component community, the maximum number of species in an infracommunity is equal to the number of species in the component community. Typically this upper limit on species richness is not realized, and no single infracommunity includes all species locally available. Among 64 component communities of intestinal parasites in eels from different localities in Britain, the maximum infracommunity richness does not exceed three parasite species per fish, whereas component communities often contain more than three species (Kennedy and Guégan, 1996). The relationship between maximum infracommunity richness and component community richness is curvilinear, such that infracommunity richness becomes increasingly independent of component community richness as the latter increases. This suggests saturation of infracommunities with parasite species at levels below the component community richness.

Saturation of infracommunities does not always occur. Among published studies on gastrointestinal helminths of birds and mammals, the relationship between maximum infracommunity richness and component community richness is linear (Figure 9.1). The slope of the relationships suggests a process of proportional sampling (Cornell and Lawton, 1992), whereby the maximum richness of infracommunities is always about one half that of the component community. Studies on species-rich helminth communities in birds suggest that empty niches are commonly observed in infracommunities (e.g. Bush and Holmes, 1986b; Stock and Holmes, 1988). Unlike the example of the eel given above, the absence of species saturation and the availability of vacant niches suggest that infracommunity richness is strictly a reflection of the availability of species for recruitment.

The richness of infracommunities can provide other clues to the existence of structure in their composition. Not all infracommunities harbour the same number of parasite species; in the same component community, some infracommunities may include only one or two species whereas others harbour over ten species. If hosts are random samplers of the parasites available in their habitat, then the richness values of infracommunities should be distributed as though species were assembled independently and randomly. If, on the other hand, assembly rules such as competitive exclusion or non-independent species recruitment are playing important roles, we would expect the distribution of infracom-

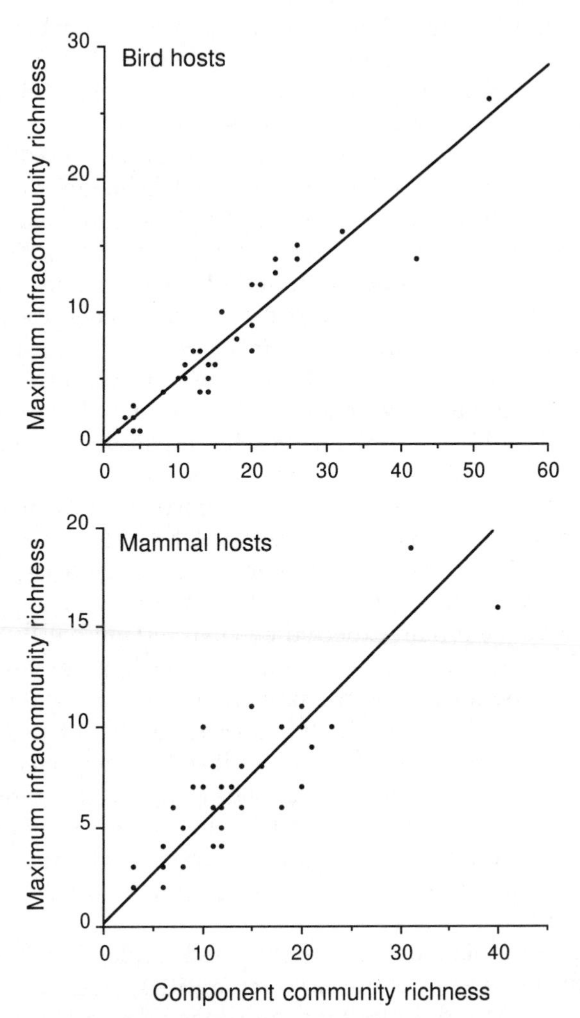

Figure 9.1 Relationship between the species richness of the component community and the maximum observed infracommunity richness, among 31 intestinal helminth communities in bird hosts and 37 in mammal hosts. For both types of host, a line fitted to the points and forced through the origin had a slope close to 0.5; functions other than linear did not provide a better fit to the points. The relationships remain linear even after log transformation of the richness values. (Data sources given in Poulin, 1996g, 1997e.)

munity richness values to differ from that expected by chance. Comparing the observed frequency distribution of infracommunity richness values with that predicted by a null model can serve as a test for the presence of structuring forces.

The Poisson distribution has been used as a null model for the distribution of infracommunity richness values (Goater *et al.*, 1987); however, it assumes equal probability of infection for all parasite species present in the component community. A better null model should account for the different species occurring with different prevalences (Adamson and Noble, 1992; Janovy *et al.*, 1995). The expected frequencies of all possible combinations of species can be computed and used to build the null distribution of infracommunity richness values (Figure 9.2). The results of published studies of parasite communities have been compared with the expectations of the null model, and the outcome suggests that random assemblages are common (Poulin, 1996g, 1997e). In about two-thirds of component communities for which data are available, the observed distribution of infracommunity richness values did not depart significantly from that predicted by the null model. This applies to intestinal parasites of all kinds of vertebrates and to ectoparasites of fish.

The cases in which the distribution of infracommunity richness values differed from that predicted by the null model may be examples of interactive communities in which species interactions cause assemblages to depart from randomness. Situations where species-poor infracommunities are more frequent than expected by chance (Figure 9.2) could be the product of strong competitive interactions such as exclusion, which would prevent the unchecked accumulation of species within infracommunities. Positive interactions among species, on the other hand, could generate more species-rich infracommunities than expected by chance, with the presence of one species facilitating the recruitment of others. Deviations from the expectations of the null model are not necessarily due to interspecific interactions, however; other explanations are possible, such as extreme heterogeneity among host individuals in susceptibility to infection.

More complex null models can also be used, models in which variables other than the prevalence of the various parasite species can be included. Janovy *et al.* (1990) used computer simulations to generate random parasite assemblages and to assess the influence of aggregation and competitive interactions. They found that the basic null model, in which the presence of one species in an infracommunity is determined by its prevalence and is independent of that of other species, provided the best fit to field data on metazoan parasites of fish.

In sharp contrast with studies of adult helminths in vertebrates, the evidence from surveys of larval digenean communities in snail intermediate hosts indicates that there are typically fewer infracommunities harbouring more than one species of digeneans than expected by chance (Figure 9.3). Interspecific competition among digeneans is the most important structuring force in these communities, with dominant species

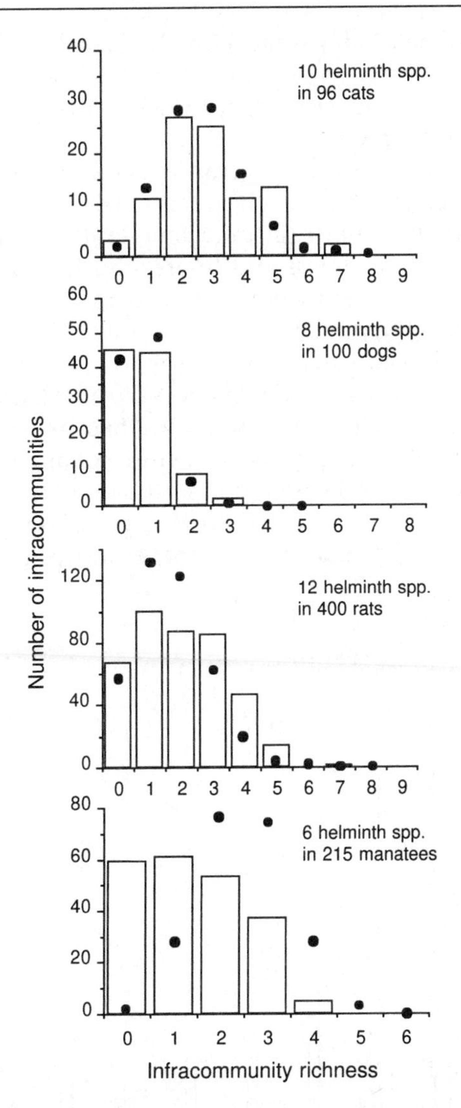

Figure 9.2 Observed (bars) and expected (black circles) frequency distributions of infracommunity richness in four component communities of gastrointestinal helminths parasitic in mammals. The expected frequencies were computed using the algorithm of Janovy *et al.* (1995) based on the respective prevalences of the different helminth species, and are what would be expected if the occurrence of species in infracommunities is independent of the presence of other species. Often observed distributions follow the expected pattern, sometimes there are more species-rich infracommunities than expected (as in rats) and sometimes there are more species-poor infracommunities than expected (as in manatees). (Data from (top to bottom) Calero *et al.*, 1951; Palmieri *et al.*, 1978; Calero *et al.*, 1950; Beck and Forrester, 1988.)

excluding subordinate ones from individual snails (Kuris and Lafferty, 1994). The different patterns presented above demonstrate that the use of null models can be a powerful tool to detect the action of structuring processes in infracommunities.

Species richness is only one descriptor of the composition of parasite infracommunities. Parasitologists have often used indices of diversity or evenness to quantify patterns in infracommunity structure. These measures combine species richness and the relative infrapopulation sizes of the different species. Diversity increases as both species richness and the evenness in their infrapopulation sizes increase. The merits of diversity and evenness indices in studies of parasite infracommunities have been questioned on statistical grounds (Simberloff and Moore, 1996). There is at least one obvious biological reason why their use can be misleading. Parasites of different species in the same infracommunity can have very different body sizes even if belonging to the same feeding guild. In the intestine of vertebrate hosts, for instance, large cestodes can dwarf other helminth parasites. In a situation like this, estimates of numerical domi-

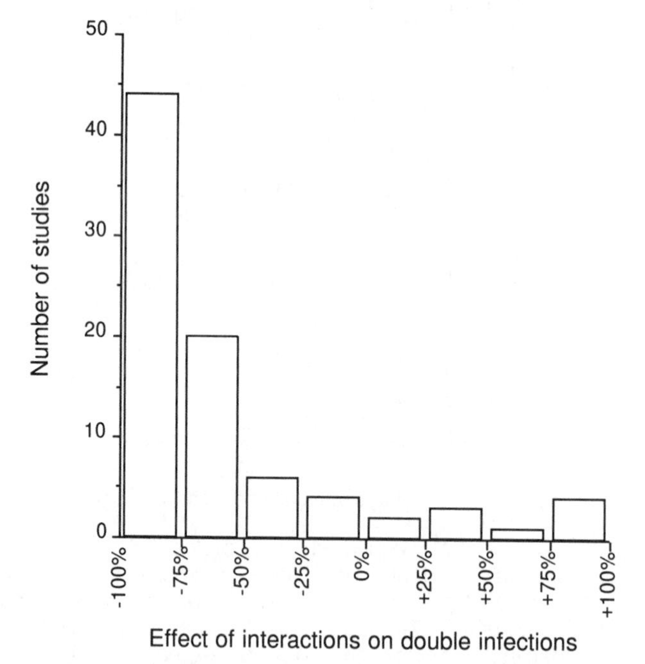

Figure 9.3 Relative effect of antagonistic interactions on the frequency of double infections of snails (i.e. infections by two parasite species) by larval digeneans across published studies in which a total of approximately 300 000 snails were examined. The effect of interactions was quantified as the relative difference between the observed and expected frequencies of double infections; a value of zero indicates no effect, and a value of –100% indicates a complete loss of double infections due to interspecific interactions. (Modified from Kuris and Lafferty, 1994.)

nance are of no importance. Measures of diversity or evenness would be more meaningful if biomass were used in the computations instead of numbers of individuals. Discrepancies in body size among helminth species in infracommunities also complicate the classification of parasites into core and satellite species, a common practice in parasite community studies (e.g. Bush and Holmes, 1986a; Esch *et al.*, 1990). Clearly, a species that is a satellite in terms of numerical abundance can be a core species in terms of biomass, making the use of relative abundance for classifying parasite species totally inadequate.

9.2 NESTED SUBSET PATTERNS IN INFRACOMMUNITIES

A non-random distribution of species richness values among infracommunities is one indication of the possible action of species interactions or other structuring forces. Another clue can be found in the species composition of infracommunities. If the species forming each infracommunity are not simply random subsets of the species available in the component community, then some species assembly rules may exist. For instance, nested subset patterns are common departures from random assemblages in free-living communities occupying insular or subdivided habitats (Patterson and Atmar, 1986; Cook, 1995; Worthen, 1996). In a parasite component community, this pattern would mean that the species forming a species-poor infracommunity are distinct subsets of progressively richer infracommunities (Figure 9.4). In other words, common parasite species, i.e. those with high prevalence, would be found in all kinds of infracommunities, but rare species would only occur in species-rich infracommunities.

What processes can generate nested subset patterns? The most likely explanation is that individual hosts are not identical habitats, and that the heterogeneity among them causes differential recruitment or extinction rates among parasite species. In communities of free-living organisms, habitat characteristics such as area and isolation can create nested patterns (Patterson and Atmar, 1986; Worthen, 1996), and host characteristics could do the same for parasite infracommunities. For example, larger-bodied host individuals may be easier to colonize because of the greater amounts of food they ingest, their larger surface area, or their greater mobility. All parasite species may colonize larger hosts, whereas only the most vagile species will encounter smaller hosts. In this scenario, larger hosts would harbour richer infracommunities than small hosts, and a nested subset pattern of parasite species could develop from small to large hosts. The size structure of the host population could thus provide the basis for the structure of parasite infracommunities.

The first test for nested patterns among parasite infracommunities was performed in a community of monogeneans ectoparasitic on a tropical freshwater fish. Guégan and Hugueny (1994) found a strong relationship

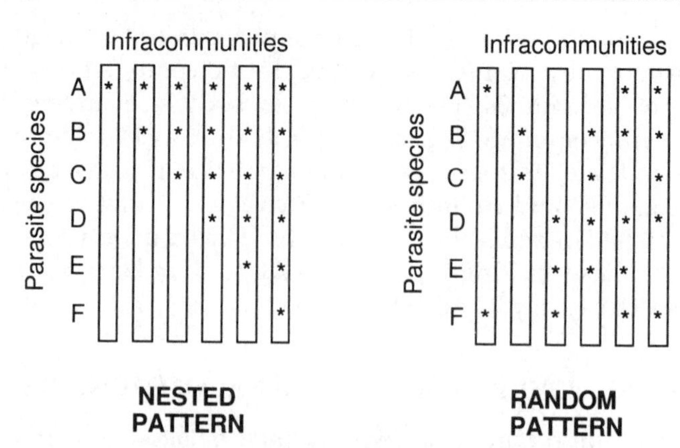

Figure 9.4 Two hypothetical distributions of parasite species among infracommunities (i.e. among individual hosts) illustrating the concept of nestedness. Each rectangle represents a different infracommunity, and infracommunities are arranged from least (at left) to most (at right) species-rich. The average infracommunity richness and the average prevalence of the six parasite species are the same for the two examples. In a perfect nested design, a parasite species occurring in a host individual with n species will be found in all host individuals with $n+1$ species.

between fish size and infracommunity richness. They also compared the observed distribution of parasite species among infracommunities to that expected under a null model of random allocation based on prevalence, and concluded that monogenean infracommunities displayed a nested species subset pattern. Guégan and Hugueny (1994) concluded that the heterogeneity among host individuals could provide a structure to parasite infracommunities. Later, Worthen and Rohde (1996) re-analysed Guégan and Hugueny's results using a more appropriate statistical procedure and found no evidence that the observed pattern differed from random assembly. In fact, Worthen and Rohde (1996) examined the component communities of metazoan ectoparasites on 38 different species of marine fish and found that nested patterns were extremely rare in these assemblages (Figure 9.5). Their results reinforce previous conclusions that fish ectoparasite communities are unstructured, random assemblages of non-interacting species.

No other type of parasite communities has received this kind of attention with regard to the existence of nested patterns. Only two component communities of intestinal helminths of mammals were studied, and neither showed a degree of nestedness greater than expected by chance (Poulin, 1996g). More empirical evidence is needed from these and other types of parasite communities. Based on the available information, however, there is no good reason to believe that infracommunities in general are something other than mere random assemblages of available species.

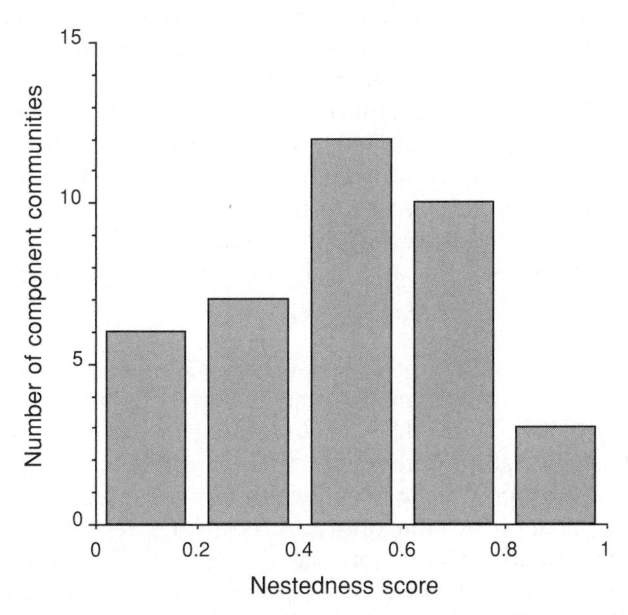

Figure 9.5 Frequency distribution of nestedness scores computed on infracommunities of 38 component communities of metazoan ectoparasites on different marine fish species. The nestedness score is the standardized index *C* of Wright and Reeves (1992), which has a maximum value of 1 when nestedness is perfect. These scores (mean ± standard deviation, 0.483 ± 0.243) do not indicate any general tendency toward nestedness in communities of fish ectoparasites. (Data from Worthen and Rohde, 1996.)

9.3 SPECIES ASSOCIATIONS AMONG INFRACOMMUNITIES

Nestedness is not the only non-random pattern one may find among a set of infracommunities. Co-occurrences of pairs or larger groups of parasite species that are more or less frequent than expected by chance could make the composition of infracommunities more or less predictable. Positive or negative associations among parasite species are perhaps the strongest suggestion that species interactions can structure infracommunities, i.e. determine their species richness and composition.

Several investigators have examined pairwise associations between parasite species across infracommunities in a component community (e.g. Moore and Simberloff, 1990; Lotz and Font, 1991; Haukisalmi and Henttonen, 1993). Associations between species can be evaluated in at least two ways. Firstly, they can be calculated using presence-absence data; a positive association found with such data implies only that two species co-occur more often than expected by chance, regardless of their numbers. Secondly, pairwise associations can be quantified as the relationship between the infection intensities of both species, or their infrapopulation sizes. Both approaches can produce spurious associations, and their interpretation must be based on comparisons with the appropriate null expectations. A third approach can be envisaged, based on the fact that numbers of parasites of different species are not always the most appropriate measures to use in tests of associations. Because different parasite species are of widely different sizes, especially among intestinal helminths, pairwise correlations between the infrapopulation biomass of two species across infracommunities can be instructive. Few investigators have used this approach (e.g. Moore and Simberloff, 1990), though it may be more sensitive and better able to detect the action of interspecific interactions.

Variance tests on binary presence-absence data such as those on parasite species in infracommunities often make the assumption that the number of positive covariances should equal the number of negative ones if parasite infracommunities are random assemblages (e.g. Schluter, 1984). An excess of positive or negative associations can be viewed as evidence for facilitation or antagonism, respectively. Several factors other than interspecific interactions can bias the sign of pairwise associations one way or the other, however, and a precise null model should account for these biases. Lotz and Font (1994) used computer simulations to demonstrate that the prevalence of two parasite species can influence the probability that a spurious association will be detected between them even if their respective occurrence in infracommunities is truly independent of one another (Figure 9.6). A rare species with low prevalence is likely to show a spurious negative association with other rare species, and spurious positive associations with common species. The rarer the species, the more pronounced the bias. The null expectation for pairwise co-occurrences involving rare species is thus not one in which positive and negative associations are equally likely.

The number of infracommunities sampled from a component community can also affect the probability of obtaining false pairwise associations between parasite species (Lotz and Font, 1994). When computing an association between two rare parasite species, a large sample size is needed before the probability of a positive covariance approaches that of a negative covariance (Figure 9.7). This is especially true for species with very low prevalence; but even for two species each with a 10% prevalence, about 100 infracommunities must be sampled before biases in the sign of spurious associations disappear. Rare species therefore introduce a bias to the analysis of co-occurrences that can only be overcome by large sample sizes.

Spurious positive or negative associations in the presence-absence data can also create false correlations between the infrapopulation sizes of pairs of parasite species. Lotz and Font (1994) quantified species associations in several component communities of intestinal helminths of bats. Using data from all infracommunities, including those in which one or

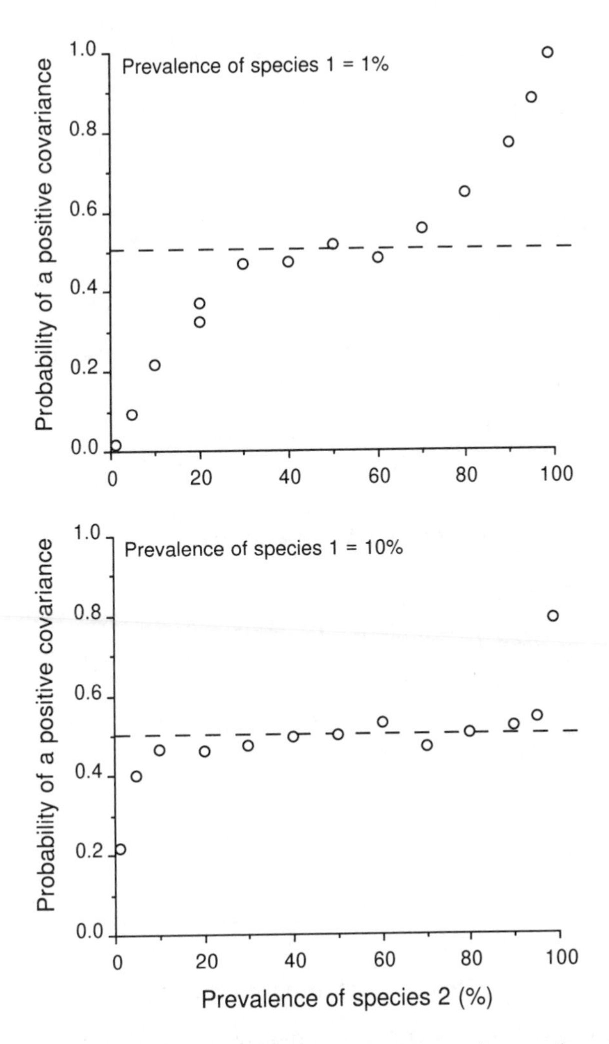

Figure 9.6 Effect of the prevalence of two parasite species on the probability of observing a positive association between them. Each point is the average of 500 Monte Carlo simulations in which sample size (the number of infracommunities sampled) was kept constant. The occurrence of each parasite species in an infracommunity was independent of the presence of the other, so that deviations from the expected value of 0.5 (broken line) represent spurious associations. (Modified from Lotz and Font, 1994.)

both members of a species pair did not occur, they found that positive associations were more frequent than expected from randomizations based on the prevalence of the species in the component communities. When only infracommunities where both members of a species pair

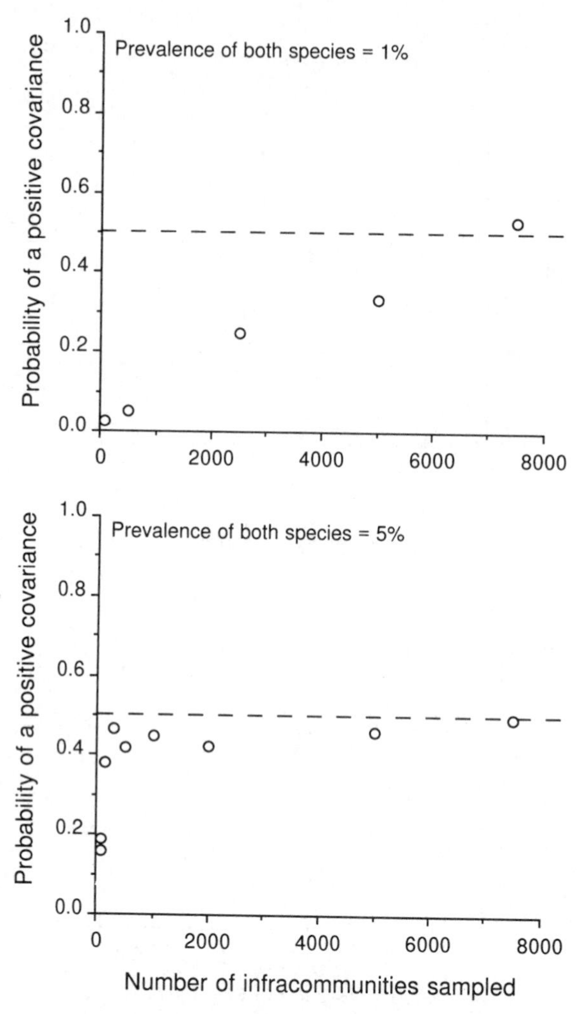

Figure 9.7 Effect of the number of infracommunities sampled on the probability of observing a positive association between two parasite species. Each point is the average of 500 Monte Carlo simulations in which the prevalence of both species was kept constant. The occurrence of each parasite species in an infracommunity was independent of the presence of the other, so that deviations from the expected value of 0.5 (broken line) represent spurious associations. (Modified from Lotz and Font, 1994.)

occur were used, however, the number of positive associations did not depart from the predicted one. These results indicate that relationships between the infrapopulation sizes of different species can be the simple consequence of joint presences or absences, and have nothing to do with interspecific interactions.

Sometimes, clear patterns emerge from an analysis of pairwise species associations. For example, Bush and Holmes (1986a) found only positive associations between all pairwise combinations among the 16 most common species of intestinal helminths in the lesser scaup duck, *Aythya affinis*. Only a quarter of these associations were statistically significant, but the fact that they all go in the same direction suggests that some force acts to provide a predictable structure to the different infracommunities. In this case, the use of common intermediate hosts by subsets of the helminth species accounted for the positive associations. Typically, however, a mixture of positive and negative associations will be observed, some of which may be statistically significant. Unless a precise null model similar to that proposed by Lotz and Font (1994) is used for comparison, the validity of emerging patterns cannot be judged. Few studies to date have used null models that take into account the biases mentioned above. At present it is therefore difficult to evaluate the generality of interspecific associations as determinants of infracommunity structure.

9.4 SPECIES RECRUITMENT AND INFRACOMMUNITY STRUCTURE

Most empirical studies of parasite infracommunities have been performed on the intestinal helminths of vertebrates, and much of the theory on parasite communities is derived from these studies. Assemblages of intestinal helminths develop from the moment an individual host acquires its first parasite, to the death of the host. An unstated assumption behind the hypothesis that interspecific interactions between parasites can structure infracommunities is that parasites join the infracommunity independently and in a random fashion, and that their initial establishment and subsequent survival are determined by the presence of other parasites, priority effects, and the strength of interspecific interactions. In fact, many intestinal helminths arrive in an infracommunity when the definitive host consumes their intermediate host. Because intermediate hosts can harbour many larval parasites of one or more species, new recruits join infracommunities in 'packets' rather than singly. One parasite species may frequently become associated with another species in the intermediate host, and the recruitment of new parasites into an infracommunity is thus a non-random process.

Bush and Holmes (1986a) found evidence that the recruitment of parasite species following different routes can have impacts on the composition and richness of infracommunities. They observed that there were two suites of helminth species, each using a different crustacean species as intermediate host, among the intestinal parasites of lesser scaup. Six cestode species and one acanthocephalan species used the amphipod *Hyalella* as intermediate host, and three cestodes and one acanthocephalan used the amphipod *Gammarus*; within each suite of species, there were consistent positive associations among species. Variability among infracommunities was to a large extent due to differences among individual hosts in the relative contribution of each amphipod to the diet.

The recruitment of new parasites into an infracommunity as packets rather than as individuals may not be limited to parasites with complex life cycles using intermediate hosts; it may extend to parasite species that release eggs in their host's faeces. Often, a faecal patch will contain eggs from different species of parasites whose adults happened to co-occur in an infracommunity. Acquisition of one parasite species by a host grazing on one patch of eggs or infective larvae may often be coupled with the acquisition of other species, so that infracommunities do not recruit new members singly and independently even if these species have simple one-host life cycles.

Intermediate hosts (or patches of infective larvae) can be seen as source communities for definitive hosts (Bush *et al.*, 1993). Infracommunities in definitive hosts may simply be the sum of non-random packets of larval parasites arriving from intermediate hosts, minus the individuals that fail to establish and survive. The structure of infracommunities in definitive hosts could represent the combined structures of the source communities acquired by the definitive host. Searching for order in the resulting assemblage may be pointless. One process which could provide a regular structure to infracommunities in definitive hosts following the stacking of source communities would be for strong interspecific interactions to override the acquired structure. Strong competition and facilitation could eliminate any patterns of association transmitted from intermediate hosts and reshape the infracommunity entirely. From the evidence presented in the preceding sections, it looks as though this does not happen often.

The frequently observed positive or negative associations between parasite species in definitive hosts could merely reflect associations existing in intermediate hosts and transferred during transmission. There may be no need to invoke interspecific interactions in the definitive host to explain these patterns. Lotz *et al.* (1995) used computer simulations to assess the contribution of species associations within intermediate hosts to the overall structure of infracommunities in definitive hosts. They found that pairwise associations between parasite species in intermediate hosts can be transferred to the infracommunities in definitive hosts (Figure 9.8). The establishment and survival rates of two associated species in the definitive host influences the extent of the association's transfer. For instance, the transfer of associations between species is nullified when the survival rate of one or both species is low because poor survival means that parasites are not transmitted as intact packets of recruits.

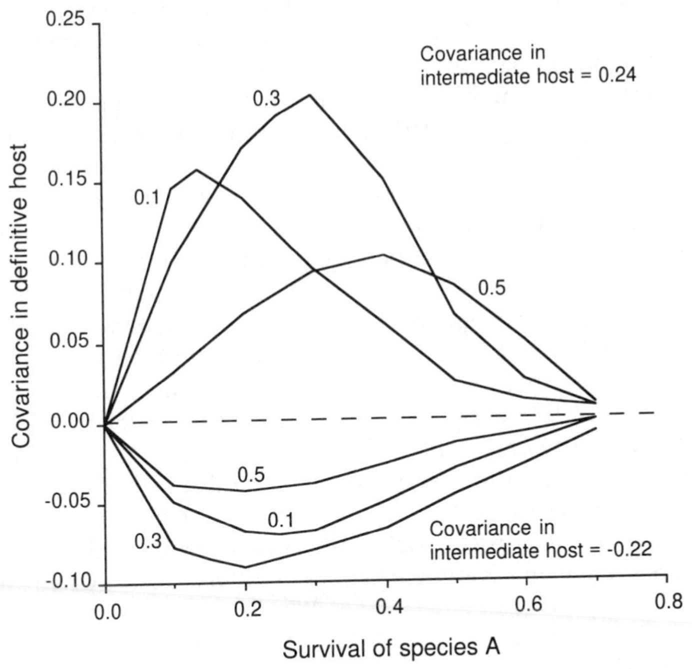

Figure 9.8 Transfer of an association between two parasite species from a source community (in intermediate hosts) to a target community (in definitive hosts). The results are from simulations in which two parasite species, A and B, share the same intermediate and definitive hosts. The two species are aggregated in the intermediate host population (k = 0.2, mean number of parasites per host = 5), and either positively or negatively associated, i.e. their infrapopulation sizes in intermediate hosts are either positively or negatively correlated. The covariance between the two species in definitive hosts is presented as a function of the survival rate of species A in the definitive host, which includes establishment success; the values on the curves represent survival of species B. The transfer of both positive (above broken line) and negative (below broken line) associations is nil when survival rates of both species are low or high, and is greatest at intermediate values when both species have similar survival rates. (Modified from Lotz *et al.*, 1995.)

In Lotz *et al.*'s (1995) simulations, the strength of the association between parasite species in definitive hosts was never as great as that in intermediate hosts (Figure 9.8). This weakening of the association is expected from the imperfect establishment and subsequent survival of parasites in their definitive host. In tests of interspecific interactions among parasites in the definitive host, associations in the intermediate hosts could serve as a rough null model specifying the maximum strength of associations in the absence of interactions. Departures from the null model, for instance stronger associations in the definitive host than in the intermediate host, could indicate the action of post-transmission processes such as facilitation. Another important result of Lotz *et al.* (1995) is that positive associations in intermediate hosts are more readily transferred to infracommunities in definitive hosts than are negative associations (Figure 9.8). This could account for the many instances in which positive associations among intestinal helminths in vertebrate hosts outnumber negative ones (e.g. Lotz and Font, 1991, 1994).

Parasite infracommunities in intermediate hosts have received very little attention, with the exception of larval digeneans in snails. Negative associations resulting from competitive exclusion or other forms of antagonism are common in these communities (Kuris and Lafferty, 1994). There is no doubt that some of that structure is transferred to infracommunities in definitive hosts. The other component of structure in these infracommunities, that produced by interspecific interactions, may mask and even completely erase the transferred structure if interactions are strong enough. The only way to distinguish between the two sources of structure would be to examine infracommunities in both intermediate and definitive hosts in the same system. There have been no such studies to date, and they may prove logistically complicated, but they are essential if we are to obtain a complete picture of infracommunity structure.

9.5 SUMMARY

Structure in a species assemblage means a predictable or repeatable departure from random assembly. It can be demonstrated only by the presence of consistent statistical associations between species or other non-random patterns. Much of this chapter has emphasized the need to contrast observed patterns with those predicted by accurate null models. This is an approach only recently adopted by parasite ecologists, and more empirical studies are needed. Based on the evidence currently available, however, it appears that the richness and composition of parasite infracommunities are often the products of stochastic events, and that infracommunities are often simply unstructured assemblages.

There have been attempts to identify the factors responsible for the development of either isolationist, unstructured communities or interactive, structured communities. Host characteristics could be important determinants of community structure. For instance, isolationist communities of intestinal helminths may be more likely in hosts with simple alimentary tracts, specialized diets, and low vagility (Kennedy *et al.*, 1986; Goater *et al.*, 1987). It has been proposed that the distinction between isolationist and interactive communities amounts to a distinction between ectothermic and endothermic hosts (Goater *et al.*, 1987). There are exceptions to this paradigm, however, that raise questions about its generality (Moore and Simberloff, 1990). Although there may be broad differences between the intestinal infracommunities of fish or amphibians on the one hand, and birds or mammals on the other, parasite communities within these host groups differ widely. The only type of parasite community for which a general conclusion is possible is the larval digeneans in their mollusc intermediate hosts. In these communities, interspecific interactions are strong enough to override the stochastic assembly of species, creating assemblages whose structure departs dramatically from randomness.

In the end, infracommunities can be understood only by examining the way in which new parasites are recruited, i.e. the way in which parasites are acquired by hosts. Recent studies have emphasized that infracommunities in definitive hosts are nothing but the combination of infracommunities from intermediate hosts, on which other processes may act. This view adds a new level to the hierarchical arrangement of parasite communities, and promotes the study of parasite communities at several levels simultaneously. There is no doubt that future studies of infracommunity structure will need to address species recruitment as well as species interactions in order to elucidate the structuring forces acting on parasite assemblages.

Component communities and parasite faunas

10

Infracommunities of parasites discussed in previous chapters are subsets of a larger assemblage of species known as the component community, or the ensemble of populations of all parasite species exploiting the host population at one point in time. Component communities are longer-lived assemblages than any of their infracommunities; they last at least a few host generations and usually much longer, as long as the host population persists in time. This creates an important distinction between the structure and dynamics of infracommunities and component communities. Whereas infracommunities are assembled over ecological time scales by infection and demographic processes, component communities are formed over evolutionary time scales by processes such as invasions, speciations, extinctions and colonizations or host switches.

The various parasite component communities of the different populations of one host species are all subsets of the entire set of parasite species exploiting that host species across its entire geographical range. This larger collection of parasite species is referred to as the parasite fauna. The term community cannot apply to the fauna because usually no single host population will harbour all species in the fauna, i.e. some species in the fauna never actually co-occur in nature. This makes parasite faunas artificial rather than biological assemblages, but the processes generating them are biological ones and worthy of discussion. Parasite faunas are formed over evolutionary time as new parasite species join component communities, and as others become extinct from all component communities. The parasite faunas of different host species will evolve over their phylogenetic history, with related species inheriting original faunas from a common ancestor which can then be modified independently in different lineages.

In this chapter, parasite assemblages are examined at higher hierarchical levels and greater spatial and temporal scales than in the previous chapter. The evolution of faunas and component communities of parasites is discussed jointly because of the clear links between these assemblages. Processes mentioned with regard to one level can easily be applied to the other. An advantage of looking at these two levels together is that we obtain a more complete picture of the evolutionary history of assemblages of parasite species.

10.1 RICHNESS AND COMPOSITION OF COMPONENT COMMUNITIES

Theoretically, the maximum number of parasite species that can be found exploiting a population of a host species equals the number of species in the parasite fauna. In practice, however, the richness of the component community rarely approaches that of the parasite fauna. There is evidence that component communities often reach a saturation level of species below that of the parasite fauna (Aho, 1990; Kennedy and Guégan, 1994, 1996). This emphasizes the artificial nature of the fauna, which is not a realistic assemblage from the parasites' perspective. Typically, in comparisons across related host species, the maximum or average richness of component communities increases with the richness of the parasite fauna until levelling off beyond a certain point (Figure 10.1). A linear relationship would indicate a process of proportional sampling, with component communities consisting of a fixed proportion of species in the fauna regardless of the richness of the fauna. The curvilinear relationship observed instead suggests that component communities become saturated with species where the curve reaches an asymptote, and that populations of host species with very rich parasite faunas do not harbour proportionately richer parasite communities than those of other host species.

The number of niches available in an individual host may set a limit on the number of parasite species that can coexist in a host population. Once the component community becomes saturated with species, further additions of species may be impossible unless new species outcompete and replace established ones. This means that the parasite species richness of the fauna is a poor predictor of the richness of component communities (Kennedy and Guégan, 1994). Except for some types of hosts or parasites, local processes are likely to be more important determinants of species richness than regional ones. In general, the physical or biological characteristics of the habitat, combined with various historical events leading to gains or losses of parasite species, will influence the local availability of parasite species and shape component communities.

Frequent contacts and exchanges of parasites between host populations of the same species should lead to highly homogeneous component communities all saturated with species. In contrast, if different host populations are isolated from one another, we may expect them to develop very different component communities. These would also tend to be very

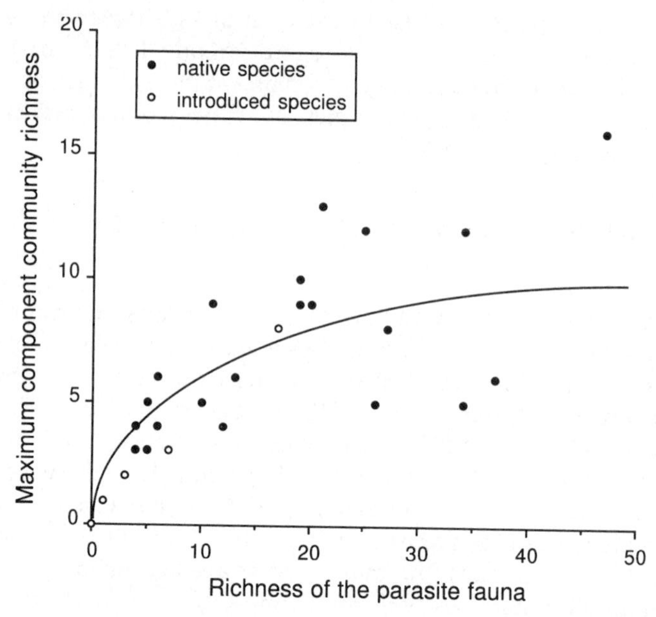

Figure 10.1 Relationship between the number of helminth parasite species in the richest observed component community and the richness of the parasite fauna, obtained from published surveys across 32 species of British freshwater fish. The curvilinear function illustrated provides the best fit to all the data points. Note, however, that a linear relationship provides the best fit to the data for introduced fish species. (Modified from Kennedy and Guégan, 1994.)

poor in species, as sampling the pool of available species in the parasite fauna would be difficult. An important trait of hosts with respect to the richness and composition of parasite communities would therefore be their vagility or migratory habits. This becomes apparent when the composition of different component communities from the same host species are compared. Parasitologists have used two ways of comparing component communities. Firstly, the Jaccard coefficient can be computed as the proportion of parasite species shared by different component communities. This gives a measure of the qualitative similarity between communities. Secondly, a more quantitative index can be used to measure the similarity of component communities based on the relative abundance of their various parasite species. Using either similarity measure, it becomes clear that component communities of host species in which populations are isolated from one another, such as freshwater fish or amphibians, are much less homogeneous than those of vagile hosts like marine fish or birds (Poulin, 1997e). The latter hosts also tend to have richer component communities, which are probably closer to saturation. Many host characteristics other than vagility may influence the richness and composition of component communities (Esch and Fernández, 1993); some of these will be addressed in section 10.3 on parasite faunas.

It is assumed in the above discussion that exchanges of parasite species between component communities are dependent only upon host movements. Exchanges can take other routes, however. Parasites may use other host species as intermediate, definitive or paratenic hosts, and use them as vehicles for the colonization of other component communities. Among species of freshwater fish, for example, parasite faunas comprising several species of helminths using birds as definitive hosts consist of more homogeneous and predictable component communities than faunas comprising mostly parasite species incapable of moving from one water body to another (Esch *et al.*, 1988). In this case, the parasites have a colonizing ability independent of the vagility of the host, which can increase the similarity of component communities and lead to their saturation in species.

Habitat characteristics are sometimes correlated with the composition or richness of component communities. These associations can sometimes be spurious, but they can also reflect the action of causative mechanisms (Hartvigsen and Halvorsen, 1994). For instance, the richness of the component communities of fish populations of the same species often correlates with selected physico-chemical characteristics of the water bodies they inhabit, such as surface area, depth, altitude or pH (e.g. Kennedy, 1978; Marcogliese and Cone, 1991, 1996). Larger water bodies are more likely to be colonized by new parasite species over time. They contain larger host populations in which parasite extinction may be less likely, and more diverse communities of free-living invertebrates that can serve as intermediate hosts to a wider variety of parasites. The pH and other chemical characteristics can determine whether essential intermediate hosts can exist in given lakes; for instance, snails and the digeneans they transmit to fish are excluded from lakes with low calcium ion concentrations (Curtis and Rau, 1980). Local habitat variables can, therefore, be more important determinants of component community richness than broader-scale factors such as the richness of the parasite fauna.

Biological characteristics of the habitats can also be important. For example, the presence of many other host species can result in a large pool of available parasite species that can lead to a richer component community. There is often a relationship between the number of host species of a given taxon in a habitat and the number of parasite species exploiting that taxon. Unionid mussels, for instance, whose larvae are ectoparasitic on fish, are more diverse in water bodies containing many fish species (Figure 10.2). This may not always lead to exchanges of parasite species between sympatric component communities, but often a larger pool of locally available species will facilitate the increase in species richness of given communities.

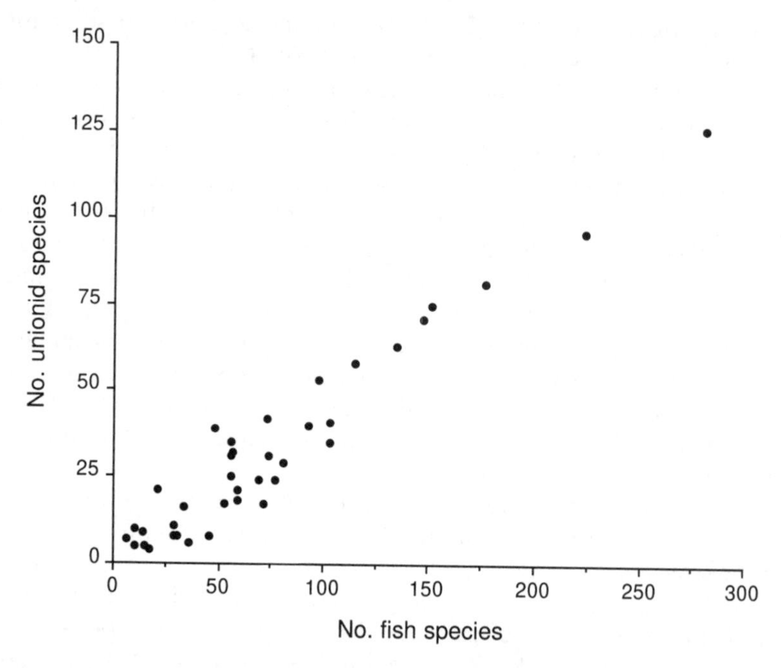

Figure 10.2 Linear relationship between the number of fish species and the number of unionid mussel species among 37 riverine systems of the Ohio River drainage area of North America. (Data from Watters, 1992.)

Even in a habitat favourable to high species richness, component communities will require time to develop to their potential. Host populations are often splintered and displaced by natural events over time. Most recently, host populations of many species have been introduced to new habitats by humans. The parasite component communities of populations immediately following these displacements are likely to be species-poor, for many reasons. For instance, essential intermediate hosts may be absent from the new habitat, or not enough parasite individuals of the same species accompany the founder host population to allow the species to maintain itself in the new habitat. The component community may re-acquire these lost parasite species over time, when new host individuals migrate into the population, and may also gain new parasite species *via* host switching from sympatric host species. The rate at which parasite species join the component community may in part be determined by the characteristics of the habitat or host species, but absolute time since the displacement can be the key factor.

The increase in component community richness over time is supported by evidence from freshwater fish in Britain. As seen in Figure 10.1, the maximum richness of component communities of helminths in introduced species increases linearly with the richness of the parasite fauna. These newly introduced fish hosts do not display the species saturation pattern characteristic of native fish species, probably because their component communities are still in a phase of species acquisition. In fact, time since the arrival of a fish species in Britain appears to be the best predictor of the richness of both its parasite fauna and component communities (Guégan and Kennedy, 1993).

The time since the displacement of a host population is not the only variable that may influence the composition or richness of component communities. The distance between the new habitat and the original habitat is also likely to be important. The farther a host population is displaced from its area of origin or its heartland, the less likely it is that its specialist parasite species will be locally available in the new habitat. For example, salmonid fish populations in their heartland harbour rich component communities comprising a high proportion of salmonid specialist parasite species (Kennedy and Bush, 1994). Many salmonid fish have been introduced to new areas, sometimes far away from their heartland. As the distance from the heartland increases, component communities become increasingly species-poor and consist of increasingly fewer salmonid specialist parasites (Figure 10.3). Salmonid specialists are gradually replaced by parasites specializing on other taxa, or by generalist parasite species.

The importance in a parasite assemblage of parasite species that are not host specialists can therefore depend on the scale of investigation. This can be seen in eels, *Anguilla rostrata*, which are more vagile than freshwater salmonids (Barker *et al.*, 1996). Component communities of eel parasites on the Atlantic coast of Canada consist mostly of eel specialist parasites. As the geographical scale of the assemblage is increased from local component communities to the entire continental fauna, the species richness increases but the relative number of specialist species decreases (Figure 10.4). The few specialist species of the fauna are well represented in most component communities, whereas component communities include only a small proportion of the large pool of non-specialist species. This illustrates that despite the relative vagility of the host species, non-specialist parasite species acquired by a host population do not necessarily spread to other component communities of the same host species.

The preceding paragraphs have emphasized that local habitat characteristics and host properties can influence the rate at which parasite species will be gained or lost by component communities over time. Two things must be pointed out, however. Firstly, stochastic events during the evolution of component communities, such as chance colonizations or extinctions, can mask any role played by the local habitat or the host species in the development of parasite assemblages (Esch *et al.*, 1988; Kennedy, 1990). Historical contingencies are also recognized as important influences during the evolution of communities of free-living organisms

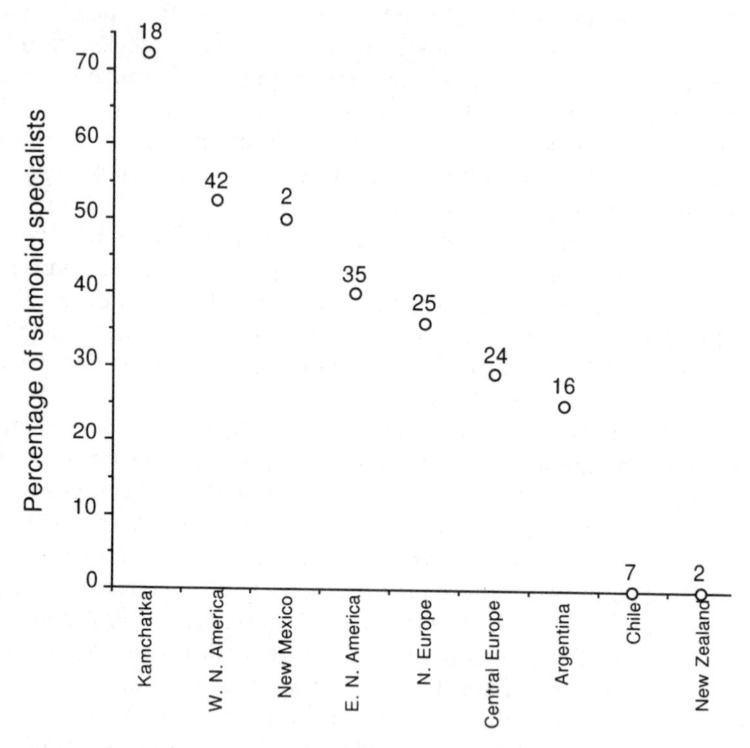

Figure 10.3 Decrease in the percentage of salmonid specialist species in helminth component communities of the rainbow trout, *Oncorhynchus mykiss*, as a function of their distance from the host species' heartland. The trout originates from the Pacific region of North America and Kamchatka; other localities are arranged in increasing distance from the heartland. Numbers above the points indicate the richness of the component communities in the various localities. (Data from Kennedy and Bush, 1994.)

(Ricklefs and Schluter, 1993). Secondly, most of the innovative tests of hypotheses regarding patterns in component community richness have been performed using data on temperate freshwater fish. Patterns observed in these hosts may not be representative of what occurs in tropical fish (e.g. Kennedy, 1995) or in groups of hosts other than fish. Similar studies are needed on component communities in other hosts to validate the trends emerging from the work on fish hosts.

10.2 EVOLUTION OF PARASITE FAUNAS

The processes described in the previous section and leading to the gradual development of component communities also shape parasite faunas. At both levels, they can be summarized as evolutionary events

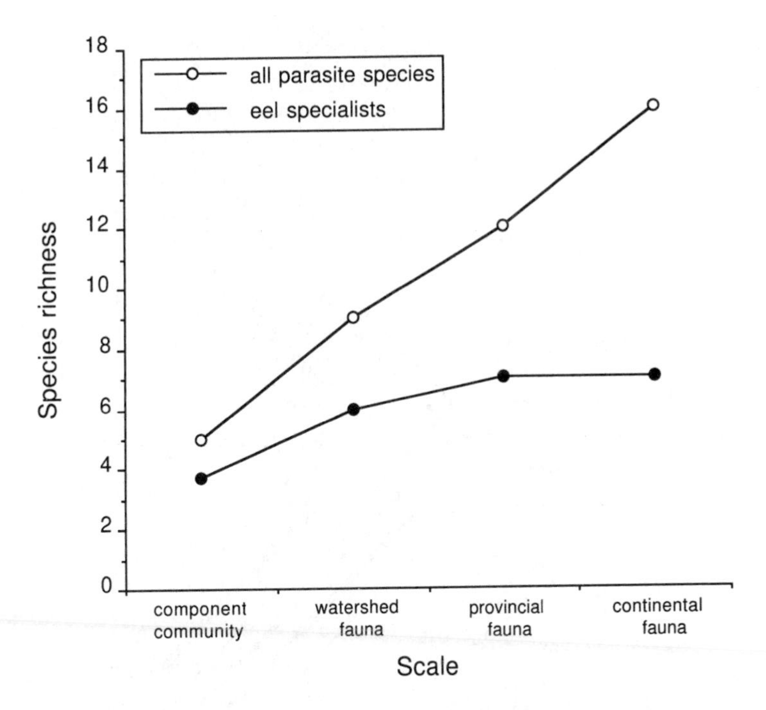

Figure 10.4 Relationship between the species richness either in all metazoan parasites or only in specialist species of eels, *Anguilla rostrata*, and the scale at which the parasite assemblage is examined. Total species richness increases linearly with scale, whereas richness in eel specialists reaches its maximum value at a geographical scale smaller than the largest one studied. (Modified from Barker *et al.*, 1996.)

leading to the acquisition or loss of parasite species (Figure 10.5). These apply equally as well to component communities in different populations of the same host species, following the breaking up of an ancestral population, as to the faunas of different host species during their phylogenetic history.

During the divergence of two parts of a host population in a host speciation event, many parasite species of the ancestral host will be inherited by the two daughter host species. These parasites may or may not cospeciate with their hosts (see Chapter 3), depending on whether the gene flow between parasites is interrupted. Cospeciation could result in different congeneric parasite species exploiting different but related host species. From a phylogenetic perspective, these congeneric parasites represent the same evolutionary lineage and cospeciation *per se* does not influence species richness. The inheritance of ancestral parasite species during host speciation creates a form of phylogenetic inertia causing related host species to have similar parasite faunas.

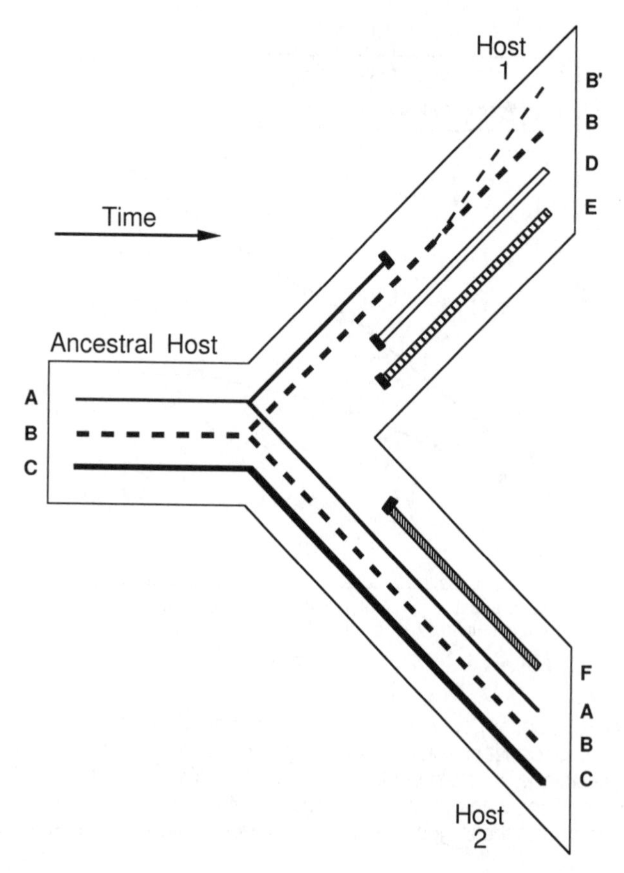

Figure 10.5 Illustration of the evolutionary events leading to the diversification of parasite component communities or faunas. Following a host speciation event, parasite lineages can be inherited by each of the two new host lineages (e.g. parasite B). They can also 'miss the boat' by being absent from the part of the host population giving rise to a new host lineage (e.g. parasite C not branching into host lineage 1), or go extinct some time after the host speciation event (e.g. parasite A in host lineage 1). New parasite lineages can also join the host lineages through colonization (e.g. parasites D, E and F) or as the result of an intrahost parasite speciation event (e.g. parasite B′ arising from parasite lineage B). (From Vickery and Poulin, 1997.)

The parasite faunas of related hosts are not identical, however. Over evolutionary time, they acquire or lose species at different rates. There are two main ways in which a fauna can lose species (Figure 10.5). Firstly, a parasite lineage may go extinct in a host lineage. Extinction of parasites can be caused by several factors. For instance, the host can evolve resistance to a particular parasite; the parasite can be displaced by a colonizing parasite species; other hosts necessary for the completion of the parasite's life cycle may disappear; or environmental changes may lead to inhospitable conditions for the free-living stages of the parasite. Secondly, because of the aggregated distribution of parasites among host individuals in a population (Chapter 6), the part of the ancestral host population splitting off to give rise to a new host species may harbour no parasites during speciation. A founder host population becoming abruptly isolated from the main body of the population may thus be free of certain parasite species, or may contain too few individual parasites to allow the parasite species to survive. The two processes described above leading to the loss of parasite species are practically indistinguishable because they differ mainly in timing. Studies of host-parasite coevolution suggest that these events are not very frequent but happen regularly (e.g. Hafner and Page, 1995; Paterson and Gray, 1996), and could potentially overcome the inertia mentioned above to create variability among the parasite faunas of related host species.

There are also two main ways in which new parasite species can be acquired by parasite faunas (Figure 10.5). Firstly, hosts can be colonized by new parasite species. These may originate from sympatric host lineages, provided that the new hosts are immunologically, physiologically and ecologically compatible with the parasites. Comparisons between reconstructed host and parasite phylogenies suggest that colonization events may have been frequent in some systems (e.g. Barker, 1991; Clayton *et al.*, 1996). In addition, sympatric host species belonging to the same taxon commonly share parasite species, suggesting either inheritance from a common host ancestor or more often the exchange of parasites among host species (Goater *et al.*, 1987; Stock and Holmes, 1987). Among four sympatric species of grebes, for instance, all but one of the 14 most common helminth species occur in more than one host species; sharing of parasite species is particularly apparent between the two fish-eating grebe species and between the two invertebrate-eating grebe species (Stock and Holmes, 1987). In other systems, however, exchanges of parasite species, or the colonization of a host species by parasites from another host species, appears to be much less frequent (e.g. Andersen and Valtonen, 1990). These exchanges are probably common only among related host species, however, as indicated by the fact that parasites of speciose host taxa often exploit a large proportion of these hosts (Poulin, 1992a).

The second way in which new parasites can be acquired by a parasite fauna involves intrahost parasite speciation (Figure 10.5). This happens when parts of the parasite population become genetically isolated without gene flow being interrupted between parts of the host population. This could coincide with new niches being developed in the host as it undergoes evolutionary changes in morphology or physiology. Intrahost parasite speciation could explain the presence of many congeneric parasite species within the same parasite fauna or even the same component community (Kennedy and Bush, 1992).

The rate at which parasite faunas acquire or lose parasite species over evolutionary time may be related to the ecological characteristics of the host species. During their phylogenetic history, related host species will diverge with respect to body size, diet, geographical range or other traits. Several studies have attempted to identify the key variables determining the richness of parasite faunas. For an effect of host ecology on the evolution of parasite faunas to be detected, however, rates of acquisition and loss of parasite species must be relatively high, and strongly linked to host ecology. This conclusion is derived from the results of computer simulations, in which an ancestral host and its parasite fauna are allowed to change through several rounds of host speciation before a correlation between host ecology and parasite fauna richness is computed (Vickery and Poulin, 1997). The inertia of parasite faunas over time is likely to be sufficient to mask the influence of host ecology, if any. The following section discusses attempts to identify influential host traits that may overcome this inertia to shape parasite faunas.

10.3 RICHNESS AND COMPOSITION OF PARASITE FAUNAS

The comparative approach is the best way to assess the role of host traits in the evolution of parasite faunas. Relating an ecological variable to the richness of the parasite fauna across host species can suggest reasons why certain host species have evolved richer faunas than others. As is apparent from the previous section, there will often be a strong phylogenetic component in parasite faunas, and the effect of ecological variables must be disentangled from that of phylogeny (Brooks, 1980; Poulin, 1995f; Gregory *et al.*, 1996). Not everyone agrees that host phylogeny is as important as, if not more important than, ecological forces (Holmes and Price, 1980; Price, 1987). The existence of comparative methods that allow the influence of phylogeny, if any, to be removed (Poulin, 1995f) makes the debate irrelevant: why not simply analyse data as though phylogeny mattered? I have no doubt that the phylogenetic component of parasite faunas is important, and strongly encourage the use of proper comparative methods in future studies.

Phylogenetic influences are not the only confounding factor requiring control in comparative analyses of parasite faunas. The other problem plaguing comparisons of parasite faunas (or of component communities, for that matter) is unequal sampling effort. As more individuals or populations of a host species are examined, the number of known parasite species in the parasite fauna increases asymptotically toward the true richness value (Figure 10.6). A parasite fauna with few known species may be truly species-poor, or it may include a large number of species yet to be recorded. As different host species have not been sampled evenly, sampling effort often explains most of the variability in richness among parasite faunas (Kuris and Blaustein, 1977; Gregory, 1990; Walther *et al.*, 1995). In fact, ecological variables often correlate closely with sampling effort. For instance, host species with large body sizes or wide geographical ranges are the subject of more independent parasite surveys than host species with small body sizes or restricted ranges. More universities are located within large geographical ranges than within small ones, and biologists show a preference toward large animals as study organisms. The close association between sampling effort and some host ecological traits make the separate effects of these variables difficult to distinguish (Guégan and Kennedy, 1996). Nevertheless, multivariate approaches or other statistical corrections must be used to discern true ecological influences from sampling artefacts (Walther *et al.*, 1995).

Keeping the potential influence of host phylogeny and sampling effort in mind, we can now review some of the proposed ecological determinants of the richness of parasite faunas. Because parasite communities are formed by colonization and extinction processes just like other communities, and because of the insular nature of hosts as habitats, MacArthur and Wilson's (1967) theory of island biogeography has been popular and influential in parasite community ecology (Kuris *et al.*, 1980; Simberloff and Moore, 1996). This theory, with others along the same lines, predicts that the number of species in a community is set by the

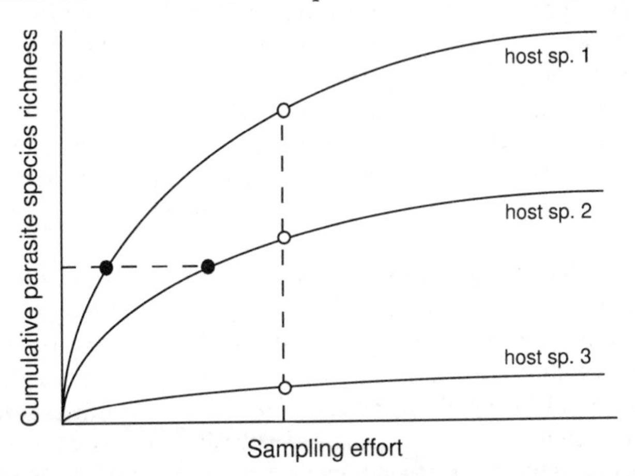

Figure 10.6 Accumulation curves for three hypothetical host species (or host populations). Sampling effort can be measured as the number of individual hosts examined for parasites, the number of parasite surveys performed on each host species, or any other measure of sampling intensity. For each host, the number of parasite species recorded approaches the true parasite species richness (the asymptote) as sampling effort increases. Typically, for a given sampling effort more parasite species will be recorded from hosts harbouring richer parasite faunas (open circles). However, when the same number of parasite species are reported from different host species (filled circles), the apparent similarity in parasite richness may be an artefact of uneven sampling effort. (Modified from Walther *et al.*, 1995.)

equilibrium between extinction and colonization rates, themselves determined by habitat characteristics. Whether the theory proves adequate or not, it has generated much empirical research and has resulted in the identification of host life-history or ecological traits associated with high parasite species richness.

In particular, host-species body size and geographical range have proven good predictors of faunal richness. The literature contains numerous examples of species-area relationships involving one or both of these traits (Price and Clancy, 1983; Gregory, 1990; Bell and Burt, 1991; Guégan *et al.*, 1992; Krasnov *et al.*, 1997). Large-bodied host species may provide more space and a broader diversity of niches for parasites. They are more likely to be colonized by parasites for these reasons, and because they consume more prey which may harbour larval parasites. They also live longer and are thus less ephemeral habitats than small-bodied, short-lived host species. Similarly, a wider geographical range may result in encounter with and colonization by a greater number of parasite species. The positive correlations between the richness of parasite faunas and either host body size or geographical range hold for a range of host and parasite types (Poulin, 1997e). It is interesting to note, however, that when correlations are computed before and after controlling for either or both host phylogeny or sampling effort, the strength of the correlations is greatly reduced after corrections are made for the confounding variables, often becoming statistically non-significant (Poulin, 1995f, 1997e). Thus, when ignoring host phylogeny, the richness of gastrointestinal helminths of birds correlates strongly with host body size; when controlling for phylogeny, the association between the two variables vanishes (Poulin, 1995f). This observation suggests that caution is needed before concluding that host ecological traits are important in shaping the evolution of parasite faunas.

Host body size and geographical range may be important in determining how many species are accumulated by parasite faunas over time, but another host trait, population density, has not received as much attention in comparative studies. A large-bodied host species occurring on all continents may be very unlikely to be colonized if its individuals are sparsely distributed within populations. Once a parasite species colonizes such a host species, it may not be able to avoid extinction, as a minimal host density is required to ensure successful transmission. Among the gastrointestinal helminth faunas of terrestrial mammals, host population density proved a better predictor of faunal richness than host body mass after the confounding influences of host phylogeny and sampling effort were removed (Morand and Poulin, 1997). The 'size' of host species as islands for colonization is thus dependent on the host biomass available to parasites, i.e. the product of host body size and density. The generality of this relationship remains to be assessed within other host taxa, however.

Ignoring host phylogeny in comparative analyses can result in a kind of pseudo-replication that artificially increases the power of statistical tests. This can lead to premature conclusions regarding the importance of certain host traits. For example, if host species are treated as independent observations, the average richness of helminth parasite faunas or component communities of aquatic vertebrates is greater than that of their terrestrial counterparts (Figure 10.7). This can suggest that habitat characteristics are the main determinant of community richness, with host species in aquatic environments encountering a greater diversity of parasites (Bush *et al.*, 1990). When phylogenetic contrasts are used instead to test for a difference in richness between aquatic and terrestrial vertebrates, no significant differences are found (Poulin, 1995f; Gregory *et al.*, 1996). Other patterns also fail to materialize once comparative analyses correct for host phylogeny. The apparent difference in species richness between the faunas of endothermic and ectothermic vertebrates, presumably due to differences in host vagility and diet (Kennedy *et al.*, 1986), cannot be demonstrated when controlling for host phylogeny.

Other host characteristics have also frequently been implicated as determinants of the richness of parasite faunas. These include host diet, behaviour and several life-history traits (Price and Clancy, 1983; Bell and Burt, 1991; Gregory *et al.*, 1991). Although these variables may indeed affect colonization or extinction rates of parasite species in faunas, more evidence is needed from comparative studies with proper corrections for phylogenetic or sampling effects. Other host traits, whose association with either parasite colonization or extinction rates is less obvious, have also been shown to correlate with the richness of parasite faunas in specific systems. For example, polyploid fish species harbour richer monogenean parasite faunas than their diploid relatives (Guégan and Morand, 1996). Several mechanisms could explain the result, but the generality of the relationship remains to be tested in other host and parasite systems.

Extrinsic factors, not related to host biology, can also create variability in the richness of parasite faunas. Among marine fish species, host species from warm regions have richer metazoan ectoparasite faunas than host species from cold regions (Poulin and Rohde, 1997). This trend remains following corrections for host phylogeny and unequal sampling effort. One possible explanation is that taxa in warm temperatures experience higher rates of diversification because shorter generation times and higher mutation rates at high temperature result in greater evolutionary speed (Rohde, 1992). Whatever the actual mechanism involved, it is clear that habitat variables can shape parasite faunas just as they influence the development of component communities.

Despite the many studies on the determinants of species richness of parasite faunas, there has been little attention paid to patterns in the composition of these faunas. Within most families of Canadian freshwater

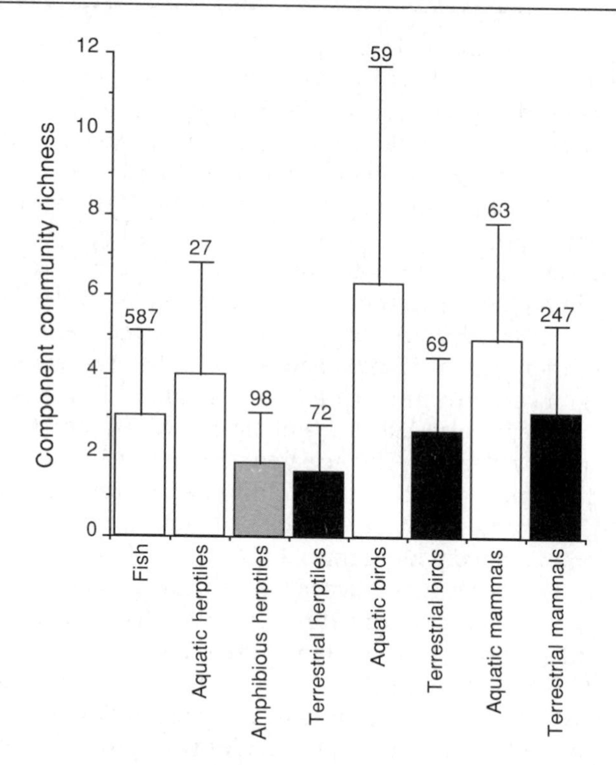

Figure 10.7 Average (± standard deviation) richness of component communities of intestinal helminth parasites in different types of vertebrate hosts. Numbers above bars show the number of studies on which the estimates are based. The data include 245 species of fish hosts, 112 species of herptile (reptile and amphibian) hosts, 84 species of bird hosts and 141 species of mammal hosts. These data suggest that aquatic vertebrates harbour richer communities of intestinal helminths than their terrestrial counterparts. (Data from Bush *et al*., 1990.)

fish, there is a significant tendency for rich parasite faunas to include many specialist, host-specific parasites, and for species-poor faunas to consist mainly of generalist parasites with low host specificity (Figure 10.8). This pattern resembles the nested pattern described for infracommunities (see Chapter 9). This non-random distribution of generalist and specialist parasite species among host species could have many explanations. A likely one could be that the proliferation of parasite species, through colonization or intrahost speciation events and low rates of extinction, is facilitated by certain host traits in some faunas whereas it is restricted in other faunas (Poulin, 1997a). The observed nested pattern fits nicely with the many studies on the determinants of the richness of parasite faunas, derived from the theory of island biogeography. Clearly, if colonization is less likely or extinctions are more frequent in some fau-

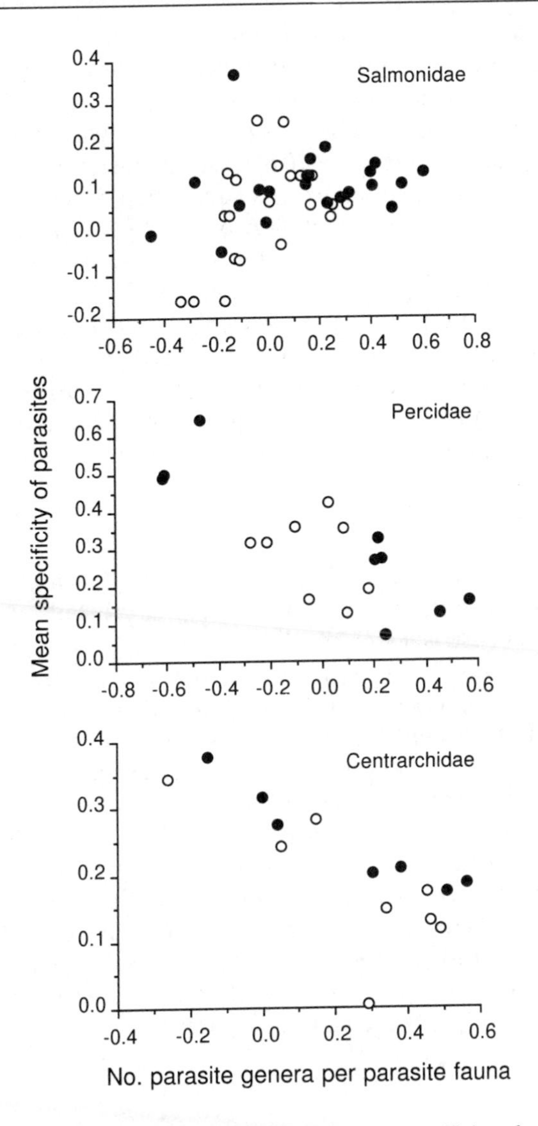

Figure 10.8 Relationship between the average host specificity of parasite species and the richness of the fauna to which they belong, across host species in three families of Canadian freshwater fish. Specificity is defined as the number of host species used by a parasite. Both specificity and faunal richness have been corrected for unequal sampling effort among fish species, i.e. data are residuals of regressions on sampling effort. Values for ectoparasite faunas (open circles) and endoparasite faunas (filled circles) are presented separately. In Percidae and Centrarchidae (as well as in Cyprinidae and Catostomidae, not shown), the relationship is negative because species-poor faunas consist mostly of generalist parasites, whereas rich faunas include many specialists. The opposite trend is observed in Salmonidae. (From Poulin, 1997a.)

nas, the few species in these faunas should also manage to colonize faunas in which conditions are more favourable. We would indeed expect generalist species to occur in all sorts of faunas and specialist ones to be restricted to rich faunas. The nested pattern was not apparent in salmonid fish, however (Figure 10.8). Fish in this family have been the subject of more introductions to new areas than other fish, which may have resulted in the recent evolution of their parasite faunas differing from that of other fish. Nested patterns are also not observed among British freshwater fish species (Guégan and Kennedy, 1996). These contradictory results suggest that more work is needed to elucidate what determines the composition of parasite faunas. As with studies of component communities, fish are the host taxon most often investigated by parasite ecologists, and information is required on the determinants of the richness and composition of parasite faunas in other groups of hosts.

10.4 SUMMARY

The development of parasite assemblages at large spatial and temporal scales proceeds through a series of evolutionary events, determined in part by the characteristics of the habitat or the host. Stochastic events such as chance colonizations no doubt play a big role, although their influence may sometimes be difficult to detect or quantify. The use of robust comparative methods for analyses of parasite assemblages is emphasized, as there is no other way of distinguishing between passive inheritance of parasite species, gain or loss of species related to host or habitat properties, and historical accidents. Many variables have been identified as determinants of high parasite species richness. However, most studies published to date have focused on the richness of helminth parasites in vertebrates, especially fish, and many have used similar data sets. Comparable studies on parasite assemblages of invertebrates would be instructive, and would serve to validate the generality of the conclusions derived from studies of vertebrate parasites.

11 Conclusion

After going through this book, many readers will feel that the common theme of the various chapters is not their evolutionary perspective of parasite ecology, but an insistence that most areas would greatly benefit from further work. It is not that we know little about parasites. The parasitology literature contains a wealth of information on the diverse taxa of parasites and on their relationship with their hosts. There are tens of thousands of pages devoted to descriptive studies, laboratory experiments and field surveys on parasites. The next step is to examine this information in a broad evolutionary context; this is where we are now. We have just begun this stage, and the evolutionary ecology of parasites is thus in its infancy. The next few years should be exciting ones as old patterns will be rigorously tested and new questions answered for the first time.

In the political climate prevailing in many countries today, funding for basic scientific research is suffering severe cutbacks. Some may therefore wonder what is the value for society of studying the evolutionary ecology of parasites. I dislike having to justify science in terms of material benefits; knowledge for its own sake should be enough. Examples of the value for society of evolutionary studies of parasites are easy to find, however. Humans have been meddling with parasite evolution for quite some time, either voluntarily or accidentally. We know that parasites can evolve rapidly in response to new selective pressures. We know this from common sense, given the short generation times and high fecundity of many parasites. We also have evidence of this rapid evolution happening, such as the sudden appearance and quick spread of anthelminthic resistance among sheep and cattle parasites throughout the world. We should therefore expect other evolutionary changes in parasites in response to the pressures we place on them.

This final chapter gives two examples of human-induced changes in the selective pressures acting on parasites, and discuss the sort of evolutionary changes they may favour. I will finish by offering my views on the immediate future of parasite evolutionary ecology, in the hope of guiding further research in what I believe are the most appropriate directions.

11.1 HABITAT CHANGES AND PARASITE EVOLUTIONARY ECOLOGY

Human activities have resulted in substantial, large-scale habitat modifications, especially in the past century. The impact of these habitat changes on parasite evolution has not received much attention but it could be very important if the new conditions persist for several parasite generations (Combes, 1995). Aquatic habitats in particular have been modified extensively in ways that may affect parasites. For instance, reservoirs created by dams and the slow-moving water of irrigation channels present ideal conditions for the proliferation of snails acting as intermediate hosts not only of human schistosomes but also of other digeneans parasitic in domestic animals or wildlife. This can increase the efficiency of transmission to the snail intermediate host and thus lower parasite mortality at the miracidial stage. Beyond an immediate increase in the parasite population, what could be the long-term or evolutionary consequences of these new conditions? We do not know. What we do know is that the evolution of life-history traits is driven mainly by an organism's age-specific mortality rate (Roff, 1992; Stearns, 1992). An increase in larval survival persisting for several parasite generations could select for a different combination of life-history traits, with changes in characteristics of the parasite such as length of the prepatent period, longevity or fecundity spreading through the population. A greater probability of transmission could also favour an increase in parasite virulence in the definitive host (see Chapter 5). At present we lack the empirical studies necessary to allow specific predictions to be made regarding these possible evolutionary changes.

Terrestrial habitats have not been spared. For example, deforestation has created patchy forest habitats separated by vast areas now used for agriculture and other human activities, or simply left vacant. Host populations in these patches may have lower densities than before deforestation, and may be more subject to local extinction. These changes may also affect parasite transmission and population dynamics. Some fear that the fragmentation of host populations will also make parasites more prone to extinction (Rózsa, 1992; Sprent, 1992). Others do not share this view and believe that many parasite attributes, evolved for other reasons, can act as protection against extinction (Bush and Kennedy, 1994). In any event, the effects of the fragmentation of habitats and host populations can influence the transmission rates and survival of larval stages, again with potential consequences on the evolution of life-history traits or transmission strategies.

Chemical pollution is perhaps one of the most pernicious and widespread environmental changes resulting from human activities. Freshwater and marine habitats are particularly susceptible, receiving industrial effluents, agricultural and domestic waste as well as many accidental spills, all containing toxic chemicals of many types. There is mounting laboratory and field evidence that these pollutants affect both parasites and their hosts (Khan and Thulin, 1991; Poulin, 1992b). These effects are not trivial: pollution levels can sometimes be accurate predictors of the presence or absence of certain parasite species in given water bodies (e.g. Marcogliese and Cone, 1996). The outcomes of pollution are varied and may depend on the host-parasite system, the habitat or the type of chemical involved. In some cases, the free-swimming infective stages of fish parasites, or the intermediate hosts of fish parasites, are negatively affected by pollutants. In other cases, fish exposed to pollutants are immunodepressed and more susceptible to parasite infections. In the first scenario larval parasite mortality is increased; in the second scenario the survival of adult parasites, as well as their growth and fecundity, is enhanced. Again, these are the sort of environmental pressures likely to lead to evolutionary changes in the biology of parasites, with the new conditions selecting for a new optimal combinations of life-history traits.

The impact of humans on the environment is far more complex than described above. Unknowingly, we may be selecting for rapid changes in parasite biology, but we do not have enough information to make specific predictions about the direction of these changes. Further work on the evolutionary ecology of parasites could provide us with the knowledge needed to forecast, and maybe plan for, the changes ahead.

11.2 PARASITE CONTROL AND PARASITE EVOLUTIONARY ECOLOGY

From a human perspective, the control strategies we use against human and livestock parasites do not often qualify as habitat changes. From the parasite perspective, however, they truly represent habitat changes causing increases in either larval or adult mortality, or both. For example, the routine use of anthelminthics against parasites of sheep and cattle has imposed new pressures on adult helminths for many parasite generations. Resistance to some anthelminthics has evolved quickly, but these have been replaced by other types of drugs so that the increased adult parasite mortality is maintained. We may thus expect other adaptive changes in parasite populations that have been exposed to anthelminthics for several years.

If the probability of adult mortality increases – in other words if the average adult life span decreases – selection may favour a shorter prepatent period, i.e. an earlier age at maturity. This would give the parasite a higher probability of producing eggs before the host is given anthelminthics. Other forms of parasite control may have different effects. Vaccines, for instance, could provide the host with immunity against parasites; if the entire host population receives the vaccine, the establishment of adult parasites is prevented and no individual has the opportunity to produce eggs. Consequences for parasite evolution would no doubt be different than those expected from the use of anthelminthics.

Other forms of parasite control target other stages in the parasite life cycle. Nematophagous fungi, which prey on the soil-dwelling infective stages of nematode parasites of livestock, are a good example. They are being touted as a promising form of biological control against these parasites (Waller, 1993). Their effect would be to increase larval nematode mortality, and one obvious evolutionary response to their prolonged action could be an increase in adult fecundity. Other control strategies aimed at larval parasites, such as the destruction of snail populations to control schistosomes, could have similar consequences.

The scenarios described above are only hypothetical, and only time and further research will verify the simple predictions made. Their purpose is only to illustrate the kinds of responses we may expect, and to show that we depend on guesswork to predict them. Evolutionary changes in parasites targeted by control strategies are not limited to livestock parasites. There is considerable evidence that the virulence of human parasites and diseases has evolved jointly with transmission routes, and that the treatment of diseases can influence virulence, by favouring more or less virulent strains (Ewald, 1994). The current awakening of medical science to the importance of evolutionary thinking in the development of long-term preventive and curative strategies (Williams and Nesse, 1991) can only help promote studies of the evolutionary ecology of parasites.

11.3 FUTURE DIRECTIONS

Recent commentaries on the future of parasitology have emphasized the need for a greater integration of the sub-disciplines of parasitology into a holistic approach to the study of parasite biology (Mettrick, 1987; Bush *et al.*, 1995). No doubt this will be necessary, especially for the comparative approach promoted here. We will require robust and highly-resolved phylogenies from molecular biologists and systematists. These are important for their own sake, but will also serve as tools for hypothesis testing by physiologists, ecologists and others. They provide the necessary evolutionary framework for studies on the biology of parasites.

The comparative approach has been applied only to some aspects of parasite biology, as will be obvious from this book. Chapters in which no comparative studies are mentioned cover subjects that are yet to benefit from this approach. There is now information available for a sufficient number of species to allow population or community level variables to be examined in the light of phylogeny, and this should be an important step in the study of parasite evolutionary ecology in the years to come.

The use of comparative studies alone is not sufficient to identify the forces shaping parasites over evolutionary time. For instance, a mechanistic approach should be used in parallel to comparative studies to identify phylogenetic constraints limiting the evolution of certain traits (McKitrick, 1993). We need studies in developmental biology or functional morphology to confirm the existence of the constraints suggested by phylogenetic analyses.

Perhaps the biggest weakness of the comparative approach is that it demonstrates an association between characters without demonstrating a causal link between them (Doughty, 1996). Over the next few years, we will need to test causal hypotheses derived from comparative studies with experiments on extant taxa (Losos, 1996). Many parasites are ideal subjects for evolutionary studies. They have small sizes, short generation times and high fecundity. They could be used in multigeneration experiments to test for the effects of particular selective regimes. An observed evolutionary change in the direction suggested by phylogenetically based comparative studies would validate the proposed causal mechanism. There may be logistical difficulties with such long-term experiments, especially if the life cycle of the parasite is not a simple one. It is also possible, however, to use natural field experiments. In the context of the habitat changes discussed above, before-and-after studies could be performed to follow the changes in parasite biology associated with a modification in their habitat such as the onset of a new control strategy, for example. These studies may not always be easy, but given the numerous opportunities that arise some may prove fruitful.

This is how I see the near future of parasite evolutionary ecology – comparative studies based on robust phylogenies used to identify associations between variables, and to generate hypotheses that can then be tested with extant taxa under experimental conditions. This research strategy will need the collaboration of people with different expertise, and more importantly it will require that scientists in the different subdisciplines of parasitology adopt a holistic, evolutionary perspective of parasite biology. There are already signs that this is happening, and the immediate future of parasite evolutionary ecology should be a bright one filled with answers.

References

Adamson, M.L. (1986) Modes of transmission and evolution of life histories in zooparasitic nematodes. *Canadian Journal of Zoology*, **64**, 1375–1384.

Adamson, M.L. and Noble, S. (1992) Structure of the pinworm (Oxyurida: Nematoda) guild in the hindgut of the American cockroach, *Periplaneta americana*. *Parasitology*, **104**, 497–507.

Adamson, M.L. and Noble, S.J. (1993) Interspecific and intraspecific competition among pinworms in the hindgut of *Periplaneta americana*. *Journal of Parasitology*, **79**, 50–56.

Adjei, E.L., Barnes, A. and Lester, R.J.G. (1986) A method for estimating possible parasite-related host mortality, illustrated using data from *Callitetrarhynchus gracilis* (Cestoda: Trypanorhyncha) in lizardfish (*Saurida* spp.). *Parasitology*, **92**, 227–243.

Adlard, R.D. and Lester, R.J.G. (1994) Dynamics of the interaction between the parasitic isopod, *Anilocra pomacentri*, and the coral reef fish, *Chromis nitida*. *Parasitology*, **109**, 311–324.

Aeby, G.S. (1992) The potential effect the ability of a coral intermediate host to regenerate has had on the evolution of its association with a marine parasite. *Proceedings of the Seventh International Coral Reef Symposium*, **2**, 809–815.

Agnew, P. and Koella, J.C. (1997) Virulence, parasite mode of transmission, and host fluctuating asymmetry. *Proceedings of the Royal Society of London B*, **264**, 9–15.

Aho, J.M. (1990) Helminth communities of amphibians and reptiles: comparative approaches to understanding patterns and processes, in *Parasite Communities: Patterns and Processes* (eds G.W. Esch, A.O. Bush and J.M. Aho), Chapman & Hall, London, pp. 157–195.

Aho, J.M. and Kennedy, C.R. (1984) Seasonal population dynamics of the nematode *Cystidicoloides tenuissima* (Zeder) from the River Swincombe, England. *Journal of Fish Biology*, **25**, 473–489.

Aho, J.M. and Kennedy, C.R. (1987) Circulation pattern and transmission dynamics of the suprapopulation of the nematode *Cystidicoloides tenuissima* (Zeder) in the River Swincombe, England. *Journal of Fish Biology*, **31**, 123–141.

Albers, G.A.A. and Grey, G.D. (1986) Breeding for worm resistance: a perspective, in *Parasitology: Quo Vadit?* Proceedings of the Sixth International Congress of Parasitology (ed. M.J. Howell), Australian Academy of Science, Canberra, pp. 559–566.

Andersen, K.I. and Valtonen, E.T. (1990) On the infracommunity structure of adult cestodes in freshwater fishes. *Parasitology*, **101**, 257–264.

Anderson, J.A., Blazek, K.J., Percival, T.J. and Janovy, J. Jr. (1993) The niche of the gill parasite *Dactylogyrus banghami* (Monogenea: Dactylogyridae) on *Notropis stramineus* (Pisces: Cyprinidae). *Journal of Parasitology*, **79**, 435–437.

Anderson, R.C. (1984) The origins of zooparasitic nematodes. *Canadian Journal of Zoology*, **62**, 317–328.

Anderson, R.C. (1996) Why do fish have so few roundworm (nematode) parasites? *Environmental Biology of Fishes*, **46**, 1–5.

Anderson, R.M. (1978) The regulation of host population growth by parasitic species. *Parasitology*, **76**, 119–157.

Anderson, R.M. (1982) Parasite dispersion patterns: generative mechanisms and dynamic consequences, in *Aspects of Parasitology* (ed. E. Meerovitch), McGill University Press, Montreal, pp. 1–40.

Anderson, R.M. (1993) Epidemiology, in *Modern Parasitology*, 2nd edn (ed. F.E.G. Cox), Blackwell, Oxford, pp. 75–116.

Anderson, R.M. (1995) Evolutionary pressures in the spread and persistence of infectious agents in vertebrate populations. *Parasitology*, **111**, S15-S31.

Anderson, R.M. and Gordon, D.M. (1982) Processes influencing the distribution of parasite numbers within host populations with special emphasis on parasite-induced host mortalities. *Parasitology*, **85**, 373–398.

Anderson, R.M. and May, R.M. (1978) Regulation and stability of host-parasite population interactions. I. Regulatory processes. *Journal of Animal Ecology*, **47**, 219–247.

Anderson, R.M. and May, R.M. (1979) Population biology of infectious diseases: part I. *Nature*, **280**, 361–367.

Anderson, R.M. and May, R.M. (1982) Coevolution of hosts and parasites. *Parasitology*, **85**, 411–426.

Anderson, R.M. and May, R.M. (1991) *Infectious Diseases of Humans*, Oxford University Press, Oxford.

Anstensrud, M. (1990) Mating strategies of two parasitic copepods [*Lernaeocera branchialis* (L.) (Pennellidae) and *Lepeophtheirus pectoralis* (Müller) (Caligidae)] on flounder: polygamy, sex-specific age at maturity and sex ratio. *Journal of Experimental Marine Biology and Ecology*, **136**, 141–158.

Athias-Binche, F. and Morand, S. (1993) From phoresy to parasitism: the example of mites and nematodes. *Research and Reviews in Parasitology*, **53**, 73–79.

Atkinson, D. (1994) Temperature and organism size: a biological law for ectotherms? *Advances in Ecological Research*, **25**, 1–58.

Ballabeni, P. (1995) Parasite-induced gigantism in a snail: a host adaptation? *Functional Ecology*, **9**, 887–893.

Ballabeni, P. and Ward, P.I. (1993) Local adaptation of the trematode *Diplostomum phoxini* to the European minnow *Phoxinus phoxinus*, its second intermediate host. *Functional Ecology*, **7**, 84–90.

Barker, D.E., Marcogliese, D.J. and Cone, D.K. (1996) On the distribution and abundance of eel parasites in Nova Scotia: local versus regional patterns. *Journal of Parasitology*, **82**, 697–701.

Barker, S.C. (1991) Evolution of host-parasite associations among species of lice and rock-wallabies: coevolution? *International Journal for Parasitology*, **21**, 497–501.

Barker, S.C. (1994) Phylogeny and classification, origins, and evolution of host associations of lice. *International Journal for Parasitology*, **24**, 1285–1291.

Barker, S.C. (1996) Lice, cospeciation and parasitism. *International Journal for Parasitology*, **26**, 219–222.

Barker, S.C. and Cribb, T.H. (1993) Sporocysts of *Mesostephanus haliasturis* (Digenea) produce miracidia. *International Journal for Parasitology*, **23**, 137–139.

Barker, S.C., Cribb, T.H., Bray, R.A. and Adlard, R.D. (1994) Host-parasite associations on a coral reef: pomacentrid fishes and digenean trematodes. *International Journal for Parasitology*, **24**, 643–647.

Barnard, C.J. (1990) Parasitic relationships, in *Parasitism and Host Behaviour* (eds C.J. Barnard and J.M. Behnke), Taylor & Francis, London, pp. 1–33.

Barral, V., Morand, S., Pointier, J.P. and Théron, A. (1996) Distribution of schistosome genetic diversity within naturally infected *Rattus rattus* detected by RAPD markers. *Parasitology*, **113**, 511–517.

Barta, J.R. (1989) Phylogenetic analysis of the class Sporozoea (phylum Apicomplexa Levine, 1970): evidence for the independent evolution of heteroxenous life cycles. *Journal of Parasitology*, **75**, 195–206.

Basch, P.F. (1990) Why do schistosomes have separate sexes? *Parasitology Today*, **6**, 160–163.

Baudoin, M. (1975) Host castration as a parasitic strategy. *Evolution*, **29**, 335–352.

Bauer, G. (1994) The adaptive value of offspring size among freshwater mussels (Bivalvia; Unionoidea). *Journal of Animal Ecology*, **63**, 933–944.

Beck, C. and Forrester, D.J. (1988) Helminths of the Florida manatee, *Trichecus manatus latirostris*, with a discussion and summary of the parasites of sirenians. *Journal of Parasitology*, **74**, 628–637.

Begon, M., Harper, J.L. and Townsend, C.R. (1996) *Ecology*, 3rd edn, Blackwell Science, Oxford.

Bell, G. and Burt, A. (1991) The comparative biology of parasite species diversity: internal helminths of freshwater fish. *Journal of Animal Ecology*, **60**, 1047–1064.

Bethel, W.M. and Holmes, J.C. (1974) Correlation of development of altered evasive behavior in *Gammarus lacustris* (Amphipoda) harboring cystacanths of *Polymorphus paradoxus* (Acanthocephala) with infectivity to the definitive host. *Journal of Parasitology*, **60**, 272–274.

Black, G.A. (1985) Reproductive output and population biology of *Cystidicola stigmatura* (Leidy) (Nematoda) in arctic char, *Salvelinus alpinus* (L.) (Salmonidae). *Canadian Journal of Zoology*, **63**, 617–622.

Black, G.A. and Lankester, M.W. (1980) Migration and development of swimbladder nematodes, *Cystidicola* spp. (Habronematoidea), in their definitive hosts. *Canadian Journal of Zoology*, **58**, 1997–2005.

Black, G.A. and Lankester, M.W. (1981) The transmission, life span, and population biology of *Cystidicola cristivomeri* White, 1941 (Nematoda: Habronematoidea) in char, *Salvelinus* spp. *Canadian Journal of Zoology*, **59**, 498–509.

Blackburn, T.M. and Gaston, K.J. (1994a) Animal body size distributions: patterns, mechanisms and implications. *Trends in Ecology & Evolution*, **9**, 471–474.

Blackburn, T.M. and Gaston, K.J. (1994b) Animal body size distributions change as more species are described. *Proceedings of the Royal Society of London B*, **257**, 293–297.

Bliss, C.I. and Fisher, R.A. (1953) Fitting the negative binomial distribution to biological data. *Biometrics*, **9**, 176–200.

Blouin, M.S., Dame, J.B., Tarrant, C.A. and Courtney, C.H. (1992) Unusual population genetics of a parasitic nematode: mtDNA variation within and among populations. *Evolution*, **46**, 470–476.

Blouin, M.S., Yowell, C.A., Courtney, C.H. and Dame, J.B. (1995) Host movement and the genetic structure of populations of parasitic nematodes. *Genetics*, **141,** 1007–1014.

Blower, S.M. and Roughgarden, J. (1989) Parasites detect host spatial pattern and density: a field experimental analysis. *Oecologia*, **78,** 138–141.

Bonhoeffer, S., Lenski, R.E. and Ebert, D. (1996) The curse of the pharaoh: the evolution of virulence in pathogens with long living propagules. *Proceedings of the Royal Society of London B*, **263,** 715–721.

Bonner, J.T. (1988) *The Evolution of Complexity by Means of Natural Selection*, Princeton University Press, Princeton.

Bonner, J.T. (1993) *Life Cycles: Reflections of an Evolutionary Biologist*, Princeton University Press, Princeton.

Bouchet, P. and Perrine, D. (1996) More gastropods feeding at night on parrot-fishes. *Bulletin of Marine Science*, **59,** 224–228.

Boulinier, T., Ives, A.R. and Danchin, E. (1996) Measuring aggregation of parasites at different host population levels. *Parasitology*, **112,** 581–587.

Brandon, R.N. (1994) Theory and experiment in evolutionary biology. *Synthese*, **99,** 59–73.

Brodeur, J. and McNeil, J.N. (1989) Seasonal microhabitat selection by an endoparasitoid through adaptive modification of host behavior. *Science*, **244,** 226–228.

Brodeur, J. and Vet, L.E.M. (1994) Usurpation of host behaviour by a parasitic wasp. *Animal Behaviour*, **48,** 187–192.

Brooks, D.R. (1980) Allopatric speciation and non-interactive parasite community structure. *Systematic Zoology*, **29,** 192–203.

Brooks, D.R. (1988) Macroevolutionary comparisons of host and parasite phylogenies. *Annual Review of Ecology and Systematics*, **19,** 235–259.

Brooks, D.R. and McLennan, D.A. (1991) *Phylogeny, Ecology, and Behavior: A Research Program in Comparative Biology*, University of Chicago Press, Chicago.

Brooks, D.R. and McLennan, D.A. (1993a) *Parascript: Parasites and the Language of Evolution*, Smithsonian Institution Press, Washington.

Brooks, D.R. and McLennan, D.A. (1993b) Comparative study of adaptive radiations with an example using parasitic flatworms (Platyhelminthes: Cercomeria). *American Naturalist*, **142,** 755–778.

Brown, A.F. (1986) Evidence for density-dependent establishment and survival of *Pomphorhynchus laevis* (Müller, 1776) (Acanthocephala) in laboratory-infected *Salmo gairdneri* Richardson and its bearing on wild populations in *Leuciscus cephalus* (L.). *Journal of Fish Biology*, **28,** 659–669.

Brusca, R.C. (1981) A monograph on the Isopoda Cymothoidae (Crustacea) of the eastern Pacific. *Zoological Journal of the Linnean Society*, **73,** 117–199.

Bull, J.J., Molineux, I.J. and Rice, W.R. (1991) Selection of benevolence in a host-parasite system. *Evolution*, **45,** 875–882.

Bullini, L., Nascetti, G., Paggi, L., Orecchia, P., Mattiucci, S. and Berland, B. (1986) Genetic variation of ascaridoid worms with different life cycles. *Evolution*, **40,** 437–440.

Bulmer, M. (1994) *Theoretical Evolutionary Ecology*, Sinauer Associates, Sunderland, Massachusetts.

Bush, A.O., Aho, J.M. and Kennedy, C.R. (1990) Ecological versus phylogenetic determinants of helminth parasite community richness. *Evolutionary Ecology*, **4,** 1–20.

Bush, A.O., Caira, J.N., Minchella, D.J., Nadler, S.A. and Seed, J.R. (1995) Parasitology year 2000. *Journal of Parasitology*, **81**, 835–842.

Bush, A.O., Heard, R.W. and Overstreet, R.M. (1993) Intermediate hosts as source communities. *Canadian Journal of Zoology*, **71**, 1358–1363.

Bush, A.O. and Holmes, J.C. (1986a) Intestinal helminths of lesser scaup ducks: patterns of association. *Canadian Journal of Zoology*, **64**, 132–141.

Bush, A.O. and Holmes, J.C. (1986b) Intestinal helminths of lesser scaup ducks: an interactive community. *Canadian Journal of Zoology*, **64**, 142–152.

Bush, A.O. and Kennedy, C.R. (1994) Host fragmentation and helminth parasites: hedging your bets against extinction. *International Journal for Parasitology*, **24**, 1333–1343.

Butterworth, E.W. and Holmes, J.C. (1984) Character divergence in two species of trematodes (*Pharyngostomoides*: Strigeoidea). *Journal of Parasitology*, **70**, 315–316.

Calero, C., Ortiz, P. and De Souza, L. (1950) Helminths in rats from Panama City and suburbs. *Journal of Parasitology*, **36**, 426.

Calero, C., Ortiz, P. and De Souza, L. (1951) Helminths in cats from Panama City and Balboa, C.Z. *Journal of Parasitology*, **37**, 326.

Calow, P. (1983) Pattern and paradox in parasite reproduction. *Parasitology*, **86** (suppl.), 197–207.

Canning, E.U. (1982) An evaluation of protozoal characteristics in relation to biological control of pests. *Parasitology*, **84**, 119–149.

Carney, J.P. and Brooks, D.R. (1991) Phylogenetic analysis of *Alloglossidium* Simer, 1929 (Digenea: Plagiorchiiformes: Macroderoididae) with discussion of the origin of truncated life cycle patterns in the genus. *Journal of Parasitology*, **77**, 890–900.

Carney, W.P. (1969) Behavioral and morphological changes in carpenter ants harboring dicrocoelid metacercariae. *American Midland Naturalist*, **82**, 605–611.

Charnov, E.L. (1982) *The Theory of Sex Allocation*, Princeton University Press, Princeton.

Chilton, N.B., Bull, C.M. and Andrews, R.H. (1992) Niche segregation in reptile ticks: attachment sites and reproductive success of females. *Oecologia*, **90**, 255–259.

Choe, J.C. and Kim, K.C. (1988) Microhabitat preference and coexistence of ectoparasitic arthropods on Alaskan seabirds. *Canadian Journal of Zoology*, **66**, 987–997.

Choe, J.C. and Kim, K.C. (1989) Microhabitat selection and coexistence in feather mites (Acari: Analgoidea) on Alaskan seabirds. *Oecologia*, **79**, 10–14.

Clarck, W.C. (1994) Origins of the parasitic habit in the Nematoda. *International Journal for Parasitology*, **24**, 1117–1129.

Clayton, D.H., Price, R.D. and Page, R.D.M. (1996) Revision of *Dennyus* (*Collodennyus*) lice (Phthiraptera: Menoponidae) from swiftlets, with descriptions of new taxa and a comparison of host-parasite relationships. *Systematic Entomology*, **21**, 179–204.

Clayton, D.H. and Tompkins, D.M. (1994) Ectoparasite virulence is linked to mode of transmission. *Proceedings of the Royal Society of London B*, **256**, 211–217.

Cockburn, A. (1991) *An Introduction to Evolutionary Ecology*, Blackwell Scientific Publications, Oxford.

Combes, C. (1990) Where do human schistosomes come from? An evolutionary approach. *Trends in Ecology & Evolution*, **5**, 334–337.

Combes, C. (1991a) Ethological aspects of parasite transmission. *American Naturalist*, **138**, 866–880.

Combes, C. (1991b) Evolution of parasite life cycles, in *Parasite-Host Associations: Coexistence or Conflict?* (eds C.A. Toft, A. Aeschlimann and L. Bolis), Oxford University Press, Oxford, pp. 62–82.

Combes, C. (1995) *Interactions Durables: Ecologie et Evolution du Parasitisme*, Masson, Paris.

Combes, C., Fournier, A., Moné, H. and Théron, A. (1994) Behaviours in trematode cercariae that enhance parasite transmission: patterns and processes. *Parasitology*, **109**, S3-S13.

Connell, J.H. (1980) Diversity and the coevolution of competitors, or the ghost of competition past. *Oikos*, **35**, 131–138.

Conway Morris, S. (1981) Parasites and the fossil record. *Parasitology*, **82**, 489–509.

Cook, R.R. (1995) The relationship between nested subsets, habitat subdivision, and species diversity. *Oecologia*, **101**, 204–210.

Cornell, H.V. and Lawton, J.H. (1992) Species interactions, local and regional processes, and limits to the richness of ecological communities: a theoretical perspective. *Journal of Animal Ecology*, **61**, 1–12.

Coustau, C., Renaud, F., Delay, B., Robbins, I. and Mathieu, M. (1991) Mechanisms involved in parasitic castration: *in vitro* effects of the trematode *Prosorhynchus squamatus* on the gametogenesis and the nutrient storage metabolism of the marine bivalve mollusc *Mytilus edulis*. *Experimental Parasitology*, **73**, 36–43.

Cox, F.E.G. (1993) *Modern Parasitology*, 2nd edn, Blackwell Scientific Publications, Oxford.

Crofton, H.D. (1971) A quantitative approach to parasitism. *Parasitology*, **62**, 179–193.

Crompton, D.W.T. (1985) Reproduction, in *Biology of the Acanthocephala* (eds D.W.T. Crompton and B.B. Nickol), Cambridge University Press, Cambridge, pp. 213–271.

Curtis, L.A. (1987) Vertical distribution of an estuarine snail altered by a parasite. *Science*, **235**, 1509–1511.

Curtis, L.A. and Hubbard, K.M.K. (1993) Species relationships in a marine gastropod-trematode ecological system. *Biological Bulletin*, **184**, 25–35.

Curtis, M.A. and Rau, M.E. (1980) The geographical distribution of diplostomiasis (Trematoda: Strigeidae) in fishes from northern Quebec, Canada, in relation to the calcium ion concentrations of lakes. *Canadian Journal of Zoology*, **58**, 1390–1394.

Dash, K.M. (1981) Interaction between *Oesophagostomum columbianum* and *Oesophagostomum venulosum* in sheep. *International Journal for Parasitology*, **11**, 201–207.

Dawkins, R. (1982) *The Extended Phenotype*, Oxford University Press, Oxford.

Dawkins, R. (1990) Parasites, desiderata lists and the paradox of the organism. *Parasitology*, **100**, S63-S73.

Day, J.F. and Edman, J.D. (1983) Malaria renders mice susceptible to mosquito feeding when gametocytes are most infective. *Journal of Parasitology*, **69**, 163–170.

De Meeüs, T., Hochberg, M.E. and Renaud, F. (1995) Maintenance of two genetic entities by habitat selection. *Evolutionary Ecology*, **9**, 131–138.

De Meeüs, T., Marin, R. and Renaud, F. (1992) Genetic heterogeneity within populations of *Lepeophtheirus europaensis* (Copepoda: Caligidae) parasitic on two host species. *International Journal for Parasitology*, **22**, 1179–1181.

De Meeüs, T., Renaud, F. and Gabrion, C. (1990) A model for studying isolation mechanisms in parasite populations: the genus *Lepeophtheirus* (Copepoda, Caligidae). *Journal of Experimental Zoology*, **254**, 207–214.

Dobson, A.P. (1985) The population dynamics of competition between parasites. *Parasitology*, **91**, 317–347.

Dobson, A.P. (1986) Inequalities in the individual reproductive success of parasites. *Parasitology*, **92**, 675–682.

Dobson, A.P. (1988) The population biology of parasite-induced changes in host behavior. *Quarterly Review of Biology*, **63**, 139–165.

Dobson, A.P. and Keymer, A.E. (1985) Life-history models, in *Biology of the Acanthocephala* (eds D.W.T. Crompton and B.B. Nickol), Cambridge University Press, Cambridge, pp. 347–384.

Dobson, A.P. and Merenlender, A. (1991) Coevolution of macroparasites and their hosts, in *Parasite-Host Associations: Coexistence or Conflict?* (eds C.A. Toft, A. Aeschlimann and L. Bolis), Oxford University Press, Oxford, pp. 83–101.

Dobson, A.P. and Roberts, M. (1994) The population dynamics of parasitic helminth communities. *Parasitology*, **109**, S97-S108.

Donnelly, R.E. and Reynolds, J.D. (1994) Occurrence and distribution of the parasitic copepod *Leposphilus labrei* on corkwing wrasse (*Crenilabrus melops*) from Mulroy Bay, Ireland. *Journal of Parasitology*, **80**, 331–332.

Doughty, P. (1996) Statistical analysis of natural experiments in evolutionary biology: comments on recent criticisms of the use of comparative methods to study adaptation. *American Naturalist*, **148**, 943–956.

Downes, B.J. (1989) Host specificity, host location and dispersal: experimental conclusions from freshwater mites (*Unionicola* spp.) parasitizing unionid mussels. *Parasitology*, **98**, 189–196.

Dufva, R. (1996) Sympatric and allopatric combinations of hen fleas and great tits: a test of the local adaptation hypothesis. *Journal of Evolutionary Biology*, **9**, 505–510.

Dunn, A.M., Adams, J. and Smith, J.E. (1993) Transovarial transmission and sex ratio distortion by a microsporidian parasite in a shrimp. *Journal of Invertebrate Pathology*, **61**, 248–252.

Dunn, A.M., Hatcher, M.J., Terry, R.S. and Tofts, C. (1995) Evolutionary ecology of vertically transmitted parasites: transovarial transmission of a microsporidian sex ratio distorter in *Gammarus duebeni*. *Parasitology*, **111**, S91-S109.

Dupont, F. and Gabrion, C. (1987) The concept of specificity in the procercoid-copepod system: *Bothriocephalus claviceps* (Cestoda) a parasite of the eel (*Anguilla anguilla*). *Parasitology Research*, **73**, 151–158.

Durette-Desset, M.-C., Beveridge, I. and Spratt, D.M. (1994) The origins and evolutionary expansion of the Strongylida (Nematoda). *International Journal for Parasitology*, **24**, 1139–1165.

Dybdahl, M.F. and Lively, C.M. (1996) The geography of coevolution: comparative population structures for a snail and its trematode parasite. *Evolution*, **50**, 2264–2275.

Ebert, D. (1994) Virulence and local adaptation of a horizontally transmitted parasite. *Science*, **265**, 1084–1086.

Ebert, D. and Herre, E.A. (1996) The evolution of parasitic diseases. *Parasitology Today*, **12**, 96–101.

Esch, G.W., Bush, A.O. and Aho, J.M. (1990) *Parasite Communities: Patterns and Processes*, Chapman & Hall, London.

Esch, G.W. and Fernández, J.C. (1993) *A Functional Biology of Parasitism: Ecological and Evolutionary Implications*, Chapman & Hall, London.

Esch, G.W., Kennedy, C.R., Bush, A.O. and Aho, J.M. (1988) Patterns in helminth communities in freshwater fish in Great Britain: alternative strategies for colonization. *Parasitology*, **96**, 519–532.

Ewald, P.W. (1983) Host-parasite relations, vectors, and the evolution of disease severity. *Annual Review of Ecology and Systematics*, **14**, 465–485.

Ewald, P.W. (1994) *Evolution of Infectious Disease*, Oxford University Press, Oxford.

Ewald, P.W. (1995) The evolution of virulence: a unifying link between parasitology and ecology. *Journal of Parasitology*, **81**, 659–669.

Frank, S.A. (1996) Models of parasite virulence. *Quarterly Review of Biology*, **71**, 37–78.

Freeland, W.J. (1983) Parasites and the coexistence of animal host species. *American Naturalist*, **121**, 223–236.

Fritz, R.S. (1982) Selection for host modification by insect parasitoids. *Evolution*, **36**, 283–288.

Fry, J.D. (1990) Trade-offs in fitness on different hosts: evidence from a selection experiment with a phytophagous mite. *American Naturalist*, **136**, 569–580.

Fulford, A.J.C. (1994) Dispersion and bias: can we trust geometric means? *Parasitology Today*, **10**, 446–448.

Garnick, E. (1992a) Niche breadth in parasites: an evolutionarily stable strategy model, with special reference to the protozoan parasite *Leishmania*. *Theoretical Population Biology*, **42**, 62–103.

Garnick, E. (1992b) Parasite virulence and parasite-host coevolution: a reappraisal. *Journal of Parasitology*, **78**, 381–386.

Gaston, K.J. (1996) Species-range-size distributions: patterns, mechanisms and implications. *Trends in Ecology & Evolution*, **11**, 197–201.

Gérard, C., Moné, H. and Théron, A. (1993) *Schistosoma mansoni-Biomphalaria glabrata*: dynamics of the sporocyst population in relation to the miracidial dose and the host size. *Canadian Journal of Zoology*, **71**, 1880–1885.

Gibson, D.I. and Bray, R.A. (1994) The evolutionary expansion and host-parasite relationships of the Digenea. *International Journal for Parasitology*, **24**, 1213–1226.

Goater, T.M., Esch G.W. and Bush, A.O. (1987) Helminth parasites of sympatric salamanders: ecological concepts at infracommunity, component and compound community levels. *American Midland Naturalist*, **118**, 289–300.

Godfray, H.C.J. (1987) The evolution of clutch size in invertebrates. *Oxford Surveys in Evolutionary Biology*, **4**, 117–154.

Godfray, H.C.J. and Werren, J.H. (1996) Recent developments in sex ratio studies. *Trends in Ecology & Evolution*, **11**, 59–63.

Gordon, D.M. and Rau, M.E. (1982) Possible evidence for mortality induced by the parasite *Apatemon gracilis* in a population of brook sticklebacks (*Culaea inconstans*). *Parasitology*, **84**, 41–47.

Gordon, D.M. and Whitfield, P.J. (1985) Interactions of the cysticercoids of *Hymenolepis diminuta* and *Raillietina cesticillus* in their intermediate host, *Tribolium confusum*. *Parasitology*, **90**, 421–431.

Grafen, A. and Woolhouse, M.E.J. (1993) Does the negative binomial distribution add up? *Parasitology Today*, **9**, 475–477.

Granath, W.O. and Esch, G.W. (1983a) Seasonal dynamics of *Bothriocephalus acheilognathi* in ambient and thermally altered areas of a North Carolina cooling reservoir. *Proceedings of the Helminthological Society of Washington*, **50**, 205–218.

Granath, W.O. and Esch, G.W. (1983b) Temperature and other factors that regulate the composition and infrapopulation densities of *Bothriocephalus acheilognathi* (Cestoda) in *Gambusia affinis* (Pisces). *Journal of Parasitology*, **69**, 1116–1124.

Gregory, R.D. (1990) Parasites and host geographic range as illustrated by waterfowl. *Functional Ecology*, **4**, 645–654.

Gregory, R.D. and Blackburn, T.M. (1991) Parasite prevalence and host sample size. *Parasitology Today*, **7**, 316–318.

Gregory, R.D., Keymer, A.E. and Harvey, P.H. (1991) Life history, ecology and parasite community structure in Soviet birds. *Biological Journal of the Linnean Society*, **43**, 249–262.

Gregory, R.D., Keymer, A.E. and Harvey, P.H. (1996) Helminth parasite richness among vertebrates. *Biodiversity and Conservation*, **5**, 985–997.

Gregory, R.D. and Woolhouse, M.E.J. (1993) Quantification of parasite aggregation: a simulation study. *Acta Tropica*, **54**, 131–139.

Guégan, J.-F. and Agnèse, J.-F. (1991) Parasite evolutionary events inferred from host phylogeny: the case of *Labeo* species (Teleostei, Cyprinidae) and their dactylogyrid parasites (Monogenea, Dactylogyridae). *Canadian Journal of Zoology*, **69**, 595–603.

Guégan, J.-F. and Hugueny, B. (1994) A nested parasite species subset pattern in tropical fish: host as major determinant of parasite infracommunity structure. *Oecologia*, **100**, 184–189.

Guégan, J.-F. and Kennedy, C.R. (1993) Maximum local helminth parasite community richness in British freshwater fish: a test of the colonization time hypothesis. *Parasitology*, **106**, 91–100.

Guégan, J.-F. and Kennedy, C.R. (1996) Parasite richness/sampling effort/host range: the fancy three-piece jigsaw puzzle. *Parasitology Today*, **12**, 367–369.

Guégan, J.-F., Lambert, A., Lévêque, C., Combes, C. and Euzet, L. (1992) Can host body size explain the parasite species richness in tropical freshwater fishes? *Oecologia*, **90**, 197–204.

Guégan, J.-F. and Morand, S. (1996) Polyploid hosts: strange attractors for parasites? *Oikos*, **77**, 366–370.

Guyatt, H.L. and Bundy, D.A.P. (1993) Estimation of intestinal nematode prevalence: influence of parasite mating patterns. *Parasitology*, **107**, 99–106.

Haas, W. (1994) Physiological analyses of host-finding behaviour in trematode cercariae: adaptations for transmission success. *Parasitology*, **109**, S15-S29.

Haas, W., Haberl, B., Kalbe, M. and Körner, M. (1995) Snail-host-finding by miracidia and cercariae: chemical host cues. *Parasitology Today*, **11**, 468–472.

Hafner, M.S. and Nadler, S.A. (1988) Phylogenetic trees support the coevolution of parasites and their hosts. *Nature*, **332**, 258–259.

Hafner, M.S. and Nadler, S.A. (1990) Cospeciation in host-parasite assemblages: comparative analysis of rates of evolution and timing of cospeciation events. *Systematic Zoology*, **39**, 192–204.

Hafner, M.S. and Page, R.D.M. (1995) Molecular phylogenies and host-parasite cospeciation: gophers and lice as a model system. *Philosophical Transactions of the Royal Society of London B*, **349**, 77–83.

Hanken, J. and Wake, D.B. (1993) Miniaturization of body size: organismal consequences and evolutionary significance. *Annual Review of Ecology and Systematics*, **24**, 501–519.

Hanski, I. (1982) Dynamics of regional distribution: the core and satellite species hypothesis. *Oikos*, **38**, 210–221.

Hartl, D.L. and Clark, A.G. (1989) *Principles of Population Genetics*, Sinauer Associates, Sunderland, Massachusetts.

Hartvigsen, R. and Halvorsen, O. (1994) Spatial patterns in the abundance and distribution of parasites of freshwater fish. *Parasitology Today*, **10**, 28–31.

Harvey, P.H. (1996) Phylogenies for ecologists. *Journal of Animal Ecology*, **65**, 255–263.

Harvey, P.H. and Keymer, A.E. (1991) Comparing life histories using phylogenies. *Philosophical Transactions of the Royal Society of London B*, **332**, 31–39.

Harvey, P.H. and Pagel, M.D. (1991) *The Comparative Method in Evolutionary Biology*, Oxford University Press, Oxford.

Hatcher, M.J. and Dunn, A.M. (1995) Evolutionary consequences of cytoplasmically inherited feminizing factors. *Philosophical Transactions of the Royal Society of London B*, **348**, 445–456.

Haukisalmi, V. and Henttonen, H. (1993) Coexistence in helminths of the bank vole, *Clethrionomys glareolus*. I. Patterns of co-occurrence. *Journal of Animal Ecology*, **62**, 221–229.

Hechtel, L.J., Johnson, C.L. and Juliano, S.A. (1993) Modification of antipredator behavior of *Caecidotea intermedius* by its parasite *Acanthocephalus dirus*. *Ecology*, **74**, 710–713.

Helluy, S. and Holmes, J.C. (1990) Serotonin, octopamine, and the clinging behavior induced by the parasite *Polymorphus paradoxus* (Acanthocephala) in *Gammarus lacustris* (Crustacea). *Canadian Journal of Zoology*, **68**, 1214–1220.

Hengeveld, R. (1992) *Dynamic Biogeography*, Cambridge University Press, Cambridge.

Hengeveld, R. and Haeck, J. (1982) The distribution of abundance. I. Measurements. *Journal of Biogeography*, **9**, 303–316.

Herre, E.A. (1993) Population structure and the evolution of virulence in nematode parasites of fig wasps. *Science*, **259**, 1442–1445.

Herre, E.A. (1995) Factors affecting the evolution of virulence: nematode parasites of fig wasps as a case study. *Parasitology*, **111**, S179-S191.

Hesselberg, C.A. and Andreassen, J. (1975) Some influences of population density on *Hymenolepis diminuta* in rats. *Parasitology*, **71**, 521–523.

Heuch, P.A. and Schram, T.A. (1996) Male mate choice in a natural population of the parasitic copepod *Lernaeocera branchialis* (Copepoda: Pennellidae). *Behaviour*, **133**, 221–239.

Hoberg, E.P, Brooks, D.R. and Siegel-Causey, D. (1996) Host-parasite cospeciation: history, principles and prospects, in *Host-Parasite Evolution: General Principles and Avian Models* (eds D.H. Clayton and J. Moore), Oxford University Press, Oxford, pp. 212–235.

Høeg, J.T. (1995) The biology and life cycle of the Rhizocephala (Cirripedia). *Journal of the Marine Biological Association U.K.*, **75**, 517–550.

Holland, C. (1984) Interactions between *Moniliformis* (Acanthocephala) and *Nippostrongylus* (Nematoda) in the small intestine of laboratory rats. *Parasitology*, **88**, 303–315.

Holmes, J.C. (1961) Effects of concurrent infections on *Hymenolepis diminuta* (Cestoda) and *Moniliformis dubius* (Acanthocephala). I. General effects and comparison with crowding. *Journal of Parasitology*, **47**, 209–216.

Holmes, J.C. (1973) Site segregation by parasitic helminths: interspecific interactions, site segregation, and their importance to the development of helminth communities. *Canadian Journal of Zoology*, **51**, 333–347.

Holmes, J.C. and Bethel, W.M. (1972) Modification of intermediate host behaviour by parasites, in *Behavioural Aspects of Parasite Transmission* (eds E.U. Canning and C.A. Wright), Academic Press, London, pp. 123–149.

Holmes, J.C. and Price, P.W. (1980) Parasite communities: the roles of phylogeny and ecology. *Systematic Zoology*, **29**, 203–213.

Holmes, J.C. and Price, P.W. (1986) Communities of parasites, in *Community Ecology: Pattern and Process* (eds D.J. Anderson and J. Kikkawa), Blackwell Scientific Publications, Oxford, pp. 187–213.

Holmes, J.C. and Zohar, S. (1990) Pathology and host behaviour, in *Parasitism and Host Behaviour* (eds C.J. Barnard and J.M. Behnke), Taylor & Francis, London, pp. 34–64.

Hoogenboom, I. and Dijkstra, C. (1987) *Sarcocystis cernae*: a parasite increasing the risk of predation of its intermediate host, *Microtus arvalis*. *Oecologia*, **74**, 86–92.

Hudson, P.J., Dobson, A.P. and Newborn, D. (1992) Do parasites make prey vulnerable to predation? Red grouse and parasites. *Journal of Animal Ecology*, **61**, 681–692.

Hurd, H. (1990) Physiological and behavioural interactions between parasites and invertebrate hosts. *Advances in Parasitology*, **29**, 271–318.

Hurd, H. and Fogo, S. (1991) Changes induced by *Hymenolepis diminuta* (Cestoda) in the behaviour of the intermediate host *Tenebrio molitor* (Coleoptera). *Canadian Journal of Zoology*, **69**, 2291–2294.

Hurst, L.D. (1993) The incidences, mechanisms and evolution of cytoplasmic sex ratio distorters in animals. *Biological Reviews*, **68**, 121–193.

Hutchinson, G.E. (1957) Concluding remarks. *Cold Spring Harbor Symposium on Quantitative Biology*, **22**, 415–427.

Huxham, M., Raffaelli, D. and Pike, A. (1993) The influence of *Cryptocotyle lingua* (Digenea: Platyhelminthes) infections on the survival and fecundity of *Littorina littorea* (Gastropoda: Prosobranchia): an ecological approach. *Journal of Experimental Marine Biology and Ecology*, **168**, 223–238.

Jablonski, D. (1996) Body size and macroevolution, in *Evolutionary Paleobiology* (eds D. Jablonski, D.H. Erwin and J.H. Lipps), University of Chicago Press, Chicago, pp. 256–289.

Jaenike, J. (1993) Rapid evolution of host specificity in a parasitic nematode. *Evolutionary Ecology*, **7**, 103–108.

Jaenike, J. (1996a) Rapid evolution of parasitic nematodes: not. *Evolutionary Ecology*, **10**, 565.

Jaenike, J. (1996b) Suboptimal virulence of an insect-parasitic nematode. *Evolution*, **50**, 2241–2247.

Jaenike, J. (1996c) Population-level consequences of parasite aggregation. *Oikos*, **76**, 155–160.

Janovy, J. Jr., Clopton, R.E., Clopton, D.A., Snyder, S.D., Efting, A. and Krebs, L. (1995) Species density distributions as null models for ecologically significant interactions of parasite species in an assemblage. *Ecological Modelling*, **77**, 189–196.

Janovy, J. Jr., Ferdig, M.T. and McDowell, M.A. (1990) A model of dynamic behavior of a parasite species assemblage. *Journal of Theoretical Biology*, **142**, 517–529.

Janovy, J. Jr. and Kutish, G.W. (1988) A model of encounters between host and parasite populations. *Journal of Theoretical Biology*, **134**, 391–401.

Jennings, J.B. and Calow, P. (1975) The relationship between high fecundity and the evolution of entoparasitism. *Oecologia*, **21**, 109–115.

Jokela, J., Uotila, L. and Taskinen, J. (1993) Effect of the castrating trematode parasite *Rhipidocotyle fennica* on energy allocation of fresh-water clam *Anondota piscinalis*. *Functional Ecology*, **7**, 332–338.

Jones, J.T., Breeze, P. and Kusel, J.R. (1989) Schistosome fecundity: influence of host genotype and intensity of infection. *International Journal for Parasitology*, **19**, 769–777.

Jourdane, J., Imbert-Establet, D. and Tchuem Tchuenté, L.A. (1995) Parthenogenesis in Schistosomatidae. *Parasitology Today*, **11**, 427–430.

Karban, R. (1989) Fine-scale adaptation of herbivorous thrips to individual host plants. *Nature*, **340**, 60–61.

Kearn, G.C. (1986) The eggs of monogeneans. *Advances in Parasitology*, **25**, 175–273.

Kennedy, C.R. (1975) *Ecological Animal Parasitology*, Blackwell Scientific Publications, Oxford.

Kennedy, C.R. (1978) An analysis of the metazoan parasitocoenoses of brown trout *Salmo trutta* from British Lakes. *Journal of Fish Biology*, **13**, 255–263.

Kennedy, C.R. (1990) Helminth communities in freshwater fish: structured communities or stochastic assemblages? in *Parasite Communities: Patterns and Processes* (eds G.W. Esch, A.O. Bush and J.M. Aho), Chapman & Hall, London, pp. 131–156.

Kennedy, C.R. (1995) Richness and diversity of macroparasite communities in tropical eels *Anguilla reinhardtii* in Queensland, Australia. *Parasitology*, **111**, 233–245.

Kennedy, C.R. and Bakke, T.A. (1989) Diversity patterns in helminth communities in common gulls, *Larus canus*. *Parasitology*, **98**, 439–445.

Kennedy, C.R. and Bush, A.O. (1992) Species richness in helminth communities: the importance of multiple congeners. *Parasitology*, **104**, 189–197.

Kennedy, C.R. and Bush, A.O. (1994) The relationship between pattern and scale in parasite communities: a stranger in a strange land. *Parasitology*, **109**, 187–196.

Kennedy, C.R., Bush, A.O. and Aho, J.M. (1986) Patterns in helminth communities: why are birds and fish so different? *Parasitology*, **93**, 205–215.

Kennedy, C.R. and Guégan, J.-F. (1994) Regional versus local helminth parasite richness in British freshwater fish: saturated or unsaturated parasite communities? *Parasitology*, **109**, 175–185.

Kennedy, C.R. and Guégan, J.-F. (1996) The number of niches in intestinal helminth communities of *Anguilla anguilla*: are there enough spaces for parasites? *Parasitology*, **113**, 293–302.

Kennedy, C.R. and Moriarty, C. (1987) Co-existence of congeneric species of Acanthocephala: *Acanthocephalus lucii* and *A. anguillae* in eels *Anguilla anguilla* in Ireland. *Parasitology*, **95**, 301–310.

Keymer, A.E. (1981) Population dynamics of *Hymenolepis diminuta* in the intermediate host. *Journal of Animal Ecology*, **50**, 941–950.

Keymer, A.E. (1982) Density-dependent mechanisms in the regulation of intestinal helminth populations. *Parasitology*, **84**, 573–587.

Keymer, A.E. and Anderson, R.M. (1979) The dynamics of infection of *Tribolium confusum* by *Hymenolepis diminuta*: the influence of infective-stage density and spatial distribution. *Parasitology*, **79**, 195–207.

Keymer, A.E., Crompton, D.W.T. and Singhvi, A. (1983) Mannose and the "crowding effect" of *Hymenolepis* in rats. *International Journal for Parasitology*, **13**, 561–570.

Keymer, A.E. and Hiorns, R.W. (1986) *Heligmosomoides polygyrus* (Nematoda): the dynamics of primary and repeated infection in outbred mice. *Proceedings of the Royal Society of London B*, **229**, 47–67.

Keymer, A.E. and Slater, A.F.G. (1987) Helminth fecundity: density dependence or statistical illusion? *Parasitology Today*, **3**, 56–58.

Khan, R.A. and Thulin, J. (1991) Influence of pollution on parasites of aquatic animals. *Advances in Parasitology*, **30**, 201–238.

Kirchner, T.B., Anderson, R.V. and Ingham, R.E. (1980) Natural selection and the distribution of nematode sizes. *Ecology*, **61**, 232–237.

Kirk, W.D.J. (1991) The size relationship between insects and their hosts. *Ecological Entomology*, **16**, 351–359.

Klompen, J.S.H., Black, W.C., Keirans, J.E. and Oliver, J.H. Jr. (1996) Evolution of ticks. *Annual Review of Entomology*, **41**, 141–161.

Koella, J.C. and Agnew, P. (1997) Blood-feeding success of the mosquito *Aedes aegypti* depends on the transmission route of its parasite *Edhazardia aedis*. *Oikos*, **78**, 311–316.

Krasnov, B.R., Shenbrot, G.I., Medvedev, S.G., Vatschenok, V.S. and Khokhlova, I.S. (1997) Host-habitat relations as an important determinant of spatial distribution of flea assemblages (Siphonaptera) on rodents in the Negev Desert. *Parasitology*, **114**, 159–173.

Kuris, A.M. (1974) Trophic interactions: similarity of parasitic castrators to parasitoids. *Quarterly Review of Biology*, **49**, 129–148.

Kuris, A.M. and Blaustein, A.R. (1977) Ectoparasitic mites on rodents: application of the island biogeography theory? *Science*, **195**, 596–598.

Kuris, A.M., Blaustein, A.R. and Alió, J.J. (1980) Hosts as islands. *American Naturalist*, **116**, 570–586.

Kuris, A.M. and Lafferty, K.D. (1994) Community structure: larval trematodes in snail hosts. *Annual Review of Ecology and Systematics*, **25**, 189–217.

Lafferty, K.D. (1992) Foraging on prey that are modified by parasites. *American Naturalist*, **140**, 854–867.

Lafferty, K.D. and Morris, A.K. (1996) Altered behavior of parasitized killifish increases susceptibility to predation by bird final hosts. *Ecology*, **77**, 1390–1397.

Lambert, A. and El Gharbi, S. (1995) Monogenean host specificity as a biological and taxonomic indicator for fish. *Biological Conservation*, **72**, 227–235.

Lawlor, B.J., Read, A.F., Keymer, A.E., Parveen, G. and Crompton, D.W.T. (1990) Non-random mating in a parasitic worm: mate choice by males? *Animal Behaviour*, **40**, 870–876.

Le Brun, N., Renaud, F., Berrebi, P. and Lambert, A. (1992) Hybrid zones and host-parasite relationships: effect on the evolution of parasitic specificity. *Evolution*, **46**, 56–61.

Lipsitch, M., Herre, E.A. and Nowak, M.A. (1995) Host population structure and the evolution of virulence: a "law of diminishing returns". *Evolution*, **49**, 743–748.

Lipsitch, M., Siller, S. and Nowak, M.A. (1996) The evolution of virulence in pathogens with vertical and horizontal transmission. *Evolution*, **50**, 1729–1741.

Lively, C.M. (1989) Adaptation by a parasitic trematode to local populations of its snail host. *Evolution*, **43**, 1663–1671.

Lloyd, M. (1967) 'Mean crowding'. *Journal of Animal Ecology*, **36**, 1–30.

Lockhart, A.B., Thrall, P.H. and Antonovics, J. (1996) Sexually transmitted diseases in animals: ecological and evolutionary implications. *Biological Reviews*, **71**, 415–471.

Loker, E.S. (1983) A comparative study of the life-histories of mammalian schistosomes. *Parasitology*, **87**, 343–369.

Losos, J.B. (1996) Phylogenies and comparative biology, stage II: testing causal hypotheses derived from phylogenies with data from extant taxa. *Systematic Biology*, **45**, 259–260.

Lotz, J.M., Bush, A.O. and Font, W.F. (1995) Recruitment-driven, spatially discontinuous communities: a null model for transferred patterns in target communities of intestinal helminths. *Journal of Parasitology*, **81**, 12–24.

Lotz, J.M. and Font, W.F. (1985) Structure of enteric helminth communities in two populations of *Eptesicus fuscus* (Chiroptera). *Canadian Journal of Zoology*, **63**, 2969–2978.

Lotz, J.M. and Font, W.F. (1991) The role of positive and negative interspecific associations in the organization of communities of intestinal helminths of bats. *Parasitology*, **103**, 127–138.

Lotz, J.M. and Font, W.F. (1994) Excess positive associations in communities of intestinal helminths of bats: a refined null hypothesis and a test of the facilitation hypothesis. *Journal of Parasitology*, **80**, 398–413.

Lowenberger, C.A. and Rau, M.E. (1994) *Plagiorchis elegans*: emergence, longevity and infectivity of cercariae, and host behavioural modifications during cercarial emergence. *Parasitology*, **109**, 65–72.

Ludwig, J.A. and Reynolds, J.F. (1988) *Statistical Ecology*, John Wiley & Sons, New York.

Lydeard, C., Mulvey, M. and Davis, G.M. (1996) Molecular systematics and evolution of reproductive traits of North American freshwater unionacean mussels (Mollusca: Bivalvia) as inferred from 16S rRNA gene sequences. *Philosophical Transactions of the Royal Society of London B*, **351**, 1593–1603.

Lymbery, A.J. (1989) Host specificity, host range and host preference. *Parasitology Today*, **5**, 298.

Lymbery, A.J. and Thompson, R.C.A. (1989) Genetic differences between cysts of *Echinococcus granulosus* from the same host. *International Journal for Parasitology*, **19**, 961–964.

Lysne, D.A., Hemmingsen, W. and Skorping, A. (1997) Regulation of infrapopulations of *Cryptocotyle lingua* on cod. *Parasitology*, **114**, 145–150.

MacArthur, R.H. and Wilson, E.O. (1967) *The Theory of Island Biogeography*, Princeton University Press, Princeton.

Maitland, D.P. (1994) A parasitic fungus infecting yellow dungflies manipulates host perching behaviour. *Proceedings of the Royal Society of London B*, **258**, 187–193.

Mangin, K.L., Lipsitch, M. and Ebert, D. (1995) Virulence and transmission modes of two microsporidia in *Daphnia magna*. *Parasitology*, **111**, 133–142.

Marcogliese, D.J. (1995) The role of zooplankton in the transmission of helminth parasites to fish. *Reviews in Fish Biology and Fisheries*, **5**, 336–371.

Marcogliese, D.J. and Cone, D.K. (1991) Importance of lake characteristics in structuring parasite communities of salmonids from insular Newfoundland. *Canadian Journal of Zoology*, **69**, 2962–2967.

Marcogliese, D.J. and Cone, D.K. (1996) On the distribution and abundance of eel parasites in Nova Scotia: influence of pH. *Journal of Parasitology*, **82**, 389–399.

May, R.M. and Anderson, R.M. (1978) Regulation and stability of host-parasite population interactions. II. Destabilizing processes. *Journal of Animal Ecology*, **47**, 249–267.

May, R.M. and Anderson, R.M. (1979) Population biology of infectious diseases: part II. *Nature*, **280**, 455–461.

May, R.M. and Woolhouse, M.E.J. (1993) Biased sex ratios and parasite mating probabilities. *Parasitology*, **107**, 287–295.

Maynard, B.J., DeMartini, L. and Wright, W.G. (1996) *Gammarus lacustris* harboring *Polymorphus paradoxus* show altered patterns of serotonin-like immunoreactivity. *Journal of Parasitology*, **82**, 663–666.

McCarthy, A.M. (1990) The influence of second intermediate host dispersion pattern upon the transmission of cercariae of *Echinoparyphium recurvatum* (Digenea: Echinostomatidae). *Parasitology*, **101**, 43–47.

McKitrick, M.C. (1993) Phylogenetic constraint in evolutionary theory: has it any explanatory power? *Annual Review of Ecology and Systematics*, **24**, 307–330.

Mettrick, D.F. (1987) Parasitology: today and tomorrow. *Canadian Journal of Zoology*, **65**, 812–822.

Minchella, D.J. (1985) Host life-history variation in response to parasitism. *Parasitology*, **90**, 205–216.

Minchella, D.J., Sollenberger, K.M. and Pereira de Souza, C. (1995) Distribution of schistosome genetic diversity within molluscan intermediate hosts. *Parasitology*, **111**, 217–220.

Mitchell, G.F., Garcia, E.G., Wood, S.M., Diasanta, R., Almonte, R., Calica, E., Davern, K.M. and Tiu, W.U. (1990) Studies on the sex ratio of worms in schistosome infections. *Parasitology*, **101**, 27–34.

Møller, A.P. (1996) Effects of host sexual selection on the population biology of parasites. *Oikos*, **75**, 340–344.

Moore, J. (1981) Asexual reproduction and environmental predictability in cestodes (Cyclophyllidea: Taeniidae). *Evolution*, **35**, 723–741.

Moore, J. (1983) Responses of an avian predator and its isopod prey to an acanthocephalan parasite. *Ecology*, **64**, 1000–1015.

Moore, J. (1984) Altered behavioral responses in intermediate hosts: an acanthocephalan parasite strategy. *American Naturalist*, **123**, 572–577.

Moore, J. (1993) Parasites and the behavior of biting flies. *Journal of Parasitology*, **79**, 1–16.

Moore, J. (1995) The behavior of parasitized animals. *BioScience*, **45**, 89–96.

Moore, J. and Brooks, D.R. (1987) Asexual reproduction in cestodes (Cyclophyllidea: Taeniidae): ecological and phylogenetic influences. *Evolution*, **41**, 882–891.

Moore, J. and Gotelli, N.J. (1990) A phylogenetic perspective on the evolution of altered host behaviours: a critical look at the manipulation hypothesis, in *Parasitism and Host Behaviour* (eds C.J. Barnard and J.M. Behnke), Taylor & Francis, London, pp. 193–233.

Moore, J. and Gotelli, N.J. (1996) Evolutionary patterns of altered behavior and susceptibility in parasitized hosts. *Evolution*, **50**, 807–819.

Moore, J. and Simberloff, D. (1990) Gastrointestinal helminth communities of bobwhite quail. *Ecology*, **71**, 344–359.

Moqbel, R. and Wakelin, D. (1979) *Trichinella spiralis* and *Strongyloides ratti*: immune interaction in adult rats. *Experimental Parasitology*, **47**, 65–72.

Morand, S. (1993) Sexual transmission of a nematode: study of a model. *Oikos*, **66**, 48–54.

Morand, S. (1996a) Biodiversity of parasites in relation to their life-cycles, in *Aspects of the Genesis and Maintenance of Biological Diversity* (eds M.E. Hochberg, J. Clobert and R. Barbault), Oxford University Press, Oxford, pp. 243–260.

Morand, S. (1996b) Life-history traits in parasitic nematodes: a comparative approach for the search of invariants. *Functional Ecology*, **10**, 210–218.

Morand, S. and Hugot, J.-P. (1997) Sexual size dimorphism in parasitic oxyuroid nematodes, in press.

Morand, S., Legendre, P., Gardner, S.L. and Hugot, J.-P. (1996a) Body size evolution of oxyurid (Nematoda) parasites: the role of hosts. *Oecologia*, **107**, 274–282.

Morand, S., Manning, S.D. and Woolhouse, M.E.J. (1996b) Parasite-host coevolution and geographic patterns of parasite infectivity and host susceptibility. *Proceedings of the Royal Society of London B*, **263**, 119–128.

Morand, S., Pointier, J.-P., Borel, G. and Théron, A. (1993) Pairing probability of schistosomes related to their distribution among the host population. *Ecology*, **74**, 2444–2449.

Morand, S. and Poulin, R. (1997) Density, body mass and parasite species richness of terrestrial mammals. *Evolutionary Ecology*, in press.

Morand, S. and Rivault, C. (1992) Infestation dynamics of *Blatticola blattae* Graeffe (Nematoda: Thelastomatidae), a parasite of *Blattella germanica* L. (Dictyoptera: Blattellidae). *International Journal for Parasitology*, **22**, 983–989.

Morand, S., Robert, F. and Connors, V.A. (1995) Complexity in parasite life cycles: population biology of cestodes in fish. *Journal of Animal Ecology*, **64**, 256–264.

Moulia, C., Le Brun, N., Dallas, J., Orth, A. and Renaud, F. (1993) Experimental evidence of genetic determinism in high susceptibility to intestinal pinworm infection in mice: a hybrid zone model. *Parasitology*, **106**, 387–393.

Mouritsen, K.N. and Jensen, K.T. (1994) The enigma of gigantism: effect of larval trematodes on growth, fecundity, egestion and locomotion in *Hydrobia ulvae* (Pennant) (Gastropoda: Prosobranchia). *Journal of Experimental Marine Biology and Ecology*, **181**, 53–66.

Mulvey, M., Aho, J.M., Lydeard, C., Leberg, P.L. and Smith, M.H. (1991) Comparative population genetic structure of a parasite (*Fascioloides magna*) and its definitive host. *Evolution*, **45**, 1628–1640.

Munger, J.C., Karasov, W.H. and Chang, D. (1989) Host genetics as a cause of overdispersion of parasites among hosts: how general a phenomenon? *Journal of Parasitology*, **75**, 707–710.

Nadler, S.A. (1990) Molecular approaches to studying helminth population genetics and phylogeny. *International Journal for Parasitology*, **20**, 11–29.

Nadler, S.A. (1995) Microevolution and the genetic structure of parasite populations. *Journal of Parasitology*, **81**, 395–403.

Nadler, S.A., Lindquist, R.L. and Near, T.J. (1995) Genetic structure of midwestern *Ascaris suum* populations: a comparison of isoenzyme and RAPD markers. *Journal of Parasitology*, **81**, 385–394.

Nevo, E. (1978) Genetic variation in natural populations: patterns and theory. *Theoretical Population Biology*, **13**, 121–177.

Nie, P. and Kennedy, C.R. (1993) Infection dynamics of larval *Bothriocephalus claviceps* in *Cyclops vicinus*. *Parasitology*, **106**, 503–509.

Noble, E.R., Noble, G.A., Schad, G.A. and MacInnes, A.J. (1989) *Parasitology: The Biology of Animal Parasites*, 6th edn, Lea & Febiger, Philadelphia.

Norval, R.A.I., Andrew, H.R. and Yunker, C.E. (1989) Pheromone-mediation of host-selection in bont ticks (*Amblyomma hebraeum* Koch). *Science*, **243**, 364–365.

Obrebski, S. (1975) Parasite reproductive strategy and evolution of castration of hosts by parasites. *Science*, **188**, 1314–1316.

Page, R.D.M. (1990) Component analysis: a valiant failure? *Cladistics*, **6**, 119–136.

Page, R.D.M. (1993) Parasites, phylogeny and cospeciation. *International Journal for Parasitology*, **23**, 499–506.

Page, R.D.M. (1994) Parallel phylogenies: reconstructing the history of host-parasite assemblages. *Cladistics*, **10**, 155–173.

Page, R.D.M., Clayton, D.H. and Paterson, A.M. (1996) Lice and cospeciation: a response to Barker. *International Journal for Parasitology*, **26**, 213–218.

Palmieri, J.R., Thurman, J.B. and Andersen, F.L. (1978) Helminth parasites of dogs in Utah. *Journal of Parasitology*, **64**, 1149–1150.

Park, T. (1948) Experimental studies of interspecies competition. I. Competition between populations of the flour beetles, *Tribolium confusum* Duval and *Tribolium castaneum* Herbst. *Ecological Monographs*, **18**, 267–307.

Partridge, L. and Harvey, P.H. (1988) The ecological context of life history evolution. *Science*, **241**, 1449–1455.

Pasternak, A.F., Huntingford, F.A. and Crompton, D.W.T. (1995) Changes in metabolism and behaviour of the freshwater copepod *Cyclops strenuus abyssorum* infected with *Diphyllobothrium* spp. *Parasitology*, **110**, 395–399.

Paterson, A.M. and Gray, R.D. (1996) Host-parasite cospeciation, host switching and missing the boat, in *Host-Parasite Evolution: General Principles and Avian Models* (eds D.H. Clayton and J. Moore), Oxford University Press, Oxford, pp. 236–250.

Paterson, A.M., Gray, R.D. and Wallis, G.P. (1993) Parasites, petrels and penguins: does louse presence reflect seabird phylogeny? *International Journal for Parasitology*, **23**, 515–526.

Patrick, M.J. (1991) Distribution of enteric helminths in *Glaucomys volans* L. (Sciuridae): a test for competition. *Ecology*, **72**, 755–758.

Patterson, B.D. and Atmar, W. (1986) Nested subsets and the structure of insular mammalian faunas and archipelagos. *Biological Journal of the Linnean Society*, **28**, 65–82.

Peters, R.H. (1983) *The Ecological Implications of Body Size*, Cambridge University Press, Cambridge.

Phares, K. (1996) An unusual host-parasite relationship: the growth hormone-like factor from plerocercoids of spirometrid tapeworms. *International Journal for Parasitology*, **26**, 575–588.

Pianka, E.R. (1970) On *r*- and *K*-selection. *American Naturalist*, **104**, 592–597.

Pianka, E.R. (1994) *Evolutionary Ecology*, 5th edn, Harper Collins, New York.

Poinar, G.O. Jr. (1983) *The Natural History of Nematodes*, Prentice Hall, Englewood Cliffs, New Jersey.

Poulin, R. (1992a) Determinants of host-specificity in parasites of freshwater fishes. *International Journal for Parasitology*, **22**, 753–758.

Poulin, R. (1992b) Toxic pollution and parasitism in freshwater fish. *Parasitology Today*, **8**, 58–61.

Poulin, R. (1993) The disparity between observed and uniform distributions: a new look at parasite aggregation. *International Journal for Parasitology*, **23**, 937–944.

Poulin, R. (1994a) The evolution of parasite manipulation of host behaviour: a theoretical analysis. *Parasitology*, **109**, S109-S118.

Poulin, R. (1994b) Meta-analysis of parasite-induced behavioural changes. *Animal Behaviour*, **48**, 137–146.

Poulin, R. (1995a) Evolutionary and ecological parasitology: a changing of the guard? *International Journal for Parasitology*, **25**, 861–862.

Poulin, R. (1995b) Clutch size and egg size in free-living and parasitic copepods: a comparative analysis. *Evolution*, **49**, 325–336.

Poulin, R. (1995c) Evolutionary influences on body size in free-living and parasitic isopods. *Biological Journal of the Linnean Society*, **54**, 231–244.

Poulin, R. (1995d) Evolution of parasite life history traits: myths and reality. *Parasitology Today*, **11**, 342–345.

Poulin, R. (1995e) "Adaptive" changes in the behaviour of parasitized animals: a critical review. *International Journal for Parasitology*, **25**, 1371–1383.

Poulin, R. (1995f) Phylogeny, ecology, and the richness of parasite communities in vertebrates. *Ecological Monographs*, **65**, 283–302.

Poulin, R. (1996a) How many parasite species are there: are we close to answers? *International Journal for Parasitology*, **26**, 1127–1129.

Poulin, R. (1996b) The evolution of life history strategies in parasitic animals. *Advances in Parasitology*, **37**, 107–134.

Poulin, R. (1996c) The evolution of body size in the Monogenea: the role of host size and latitude. *Canadian Journal of Zoology*, **74**, 726–732.

Poulin, R. (1996d) Sexual size dimorphism and transition to parasitism in copepods. *Evolution*, **50**, 2520–2523.

Poulin, R. (1996e) Observations on the free-living adult stage of *Gordius dimorphus* (Nematomorpha: Gordioidea). *Journal of Parasitology*, **82**, 845–846.

Poulin, R. (1996f) Measuring parasite aggregation: defending the index of discrepancy. *International Journal for Parasitology*, **26**, 227–229.

Poulin, R. (1996g) Richness, nestedness, and randomness in parasite infracommunity structure. *Oecologia*, **105**, 545–551.

Poulin, R. (1997a) Parasite faunas of freshwater fish: the relationship between richness and the specificity of parasites. *International Journal for Parasitology*, in press.

Poulin, R. (1997b) Egg production in adult trematodes: adaptation or constraint? *Parasitology*, **114**, 195–204.

Poulin, R. (1997c) Host and environmental correlates of body size in ticks (Acari: Ixodida), in press.

Poulin, R. (1997d) Population abundance and sex ratio in dioecious helminth parasites. *Oecologia*, in press.

Poulin, R. (1997e) Species richness of parasite assemblages: evolution and patterns. *Annual Review of Ecology and Systematics*, in press.

Poulin, R., Curtis, M.A. and Rau, M.E. (1992) Effects of *Eubothrium salvelini* (Cestoda) on the behaviour of *Cyclops vernalis* (Copepoda) and its susceptibility to fish predators. *Parasitology*, **105**, 265–271.

Poulin, R. and Hamilton, W.J. (1995) Ecological determinants of body size and clutch size in amphipods: a comparative approach. *Functional Ecology*, **9**, 364–370.

Poulin, R. and Hamilton, W.J. (1997) Ecological correlates of body size and egg size in parasitic Ascothoracida and Rhizocephala (Crustacea). *Acta Oecologica*, in press.

Poulin, R. and Morand, S. (1997) Parasite body size distributions: interpreting patterns of skewness. *International Journal for Parasitology*, in press.

Poulin, R., Rau, M.E. and Curtis, M.A. (1991) Infection of brook trout fry, *Salvelinus fontinalis*, by ectoparasitic copepods: the role of host behaviour and initial parasite load. *Animal Behaviour*, **41**, 467–476.

Poulin, R. and Rohde, K. (1997) Comparing the richness of metazoan ectoparasite communities of marine fishes: controlling for host phylogeny. *Oecologia*, **110**, 278–283.

Price, P.W. (1974) Strategies for egg production. *Evolution*, **28**, 76–84.

Price, P.W. (1980) *Evolutionary Biology of Parasites*, Princeton University Press, Princeton.

Price, P.W. (1987) Evolution in parasite communities. *International Journal for Parasitology*, **17**, 209–214.

Price, P.W. and Clancy, K.M. (1983) Patterns in number of helminth parasite species in freshwater fishes. *Journal of Parasitology*, **69**, 449–454.

Quinnell, R.J., Medley, G.F. and Keymer, A.E. (1990) The regulation of gastrointestinal helminth populations. *Philosophical Transactions of the Royal Society of London B*, **330**, 191–201.

Raibaut, A. and Trilles, J.P. (1993) The sexuality of parasitic crustaceans. *Advances in Parasitology*, **32**, 367–444.

Read, A.F., Anwar, M., Shutler, D. and Nee, S. (1995) Sex allocation and population structure in malaria and related parasitic protozoa. *Proceedings of the Royal Society of London B*, **260**, 359–363.

Read, A.F., Narara, A., Nee, S., Keymer, A.E. and Day, K.P. (1992) Gametocyte sex ratios as indirect measures of outcrossing rates in malaria. *Parasitology*, **104**, 387–395.

Read, A.F. and Skorping, A. (1995) The evolution of tissue migration by parasitic nematode larvae. *Parasitology*, **111**, 359–371.

Read, C.P. (1951) The "crowding effect" in tapeworm infections. *Journal of Parasitology*, **37**, 174–178.

Rékási, J., Rózsa, L. and Kiss, B.J. (1997) Patterns in the distribution of avian lice (Phthiraptera: Amblycera, Ischnocera). *Journal of Avian Biology*, **28**, 150–156.

Reversat, J., Silan, P. and Maillard, C. (1992) Structure of monogenean populations, ectoparasites of the gilthead sea bream *Sparus aurata*. *Marine Biology*, **112**, 43–47.

Ricklefs, R.E. and Schluter, D. (1993) *Species Diversity in Ecological Communities: Historical and Geographical Perspectives*, University of Chicago Press, Chicago.

Riggs, M.R. and Esch, G.W. (1987) The suprapopulation dynamics of *Bothriocephalus acheilognathi* in a North Carolina reservoir: abundance, dispersion, and prevalence. *Journal of Parasitology*, **73**, 877–892.

Riggs, M.R., Lemly, A.D. and Esch, G.W. (1987) The growth, biomass, and fecundity of *Bothriocephalus acheilognathi* in a North Carolina cooling reservoir. *Journal of Parasitology*, **73**, 893–900.

Robb, T. and Reid, M.L. (1996) Parasite-induced changes in the behaviour of cestode-infected beetles: adaptation or simple pathology? *Canadian Journal of Zoology*, **74**, 1268–1274.

Robert, F., Renaud, F., Mathieu, E. and Gabrion, C. (1988) Importance of the paratenic host in the biology of *Bothriocephalus gregarius* (Cestoda, Pseudophyllidea), a parasite of the turbot. *International Journal for Parasitology*, **18**, 611–621.

Roche, M. and Patrzek, D. (1966) The female to male ratio (FMR) in hookworm. *Journal of Parasitology*, **52**, 117–121.

Roff, D.A. (1992) *The Evolution of Life Histories: Theory and Analysis*, Chapman & Hall, New York.

Rohde, K. (1979) A critical evaluation of intrinsic and extrinsic factors responsible for niche restriction in parasites. *American Naturalist*, **114**, 648–671.

Rohde, K. (1980) Host specificity indices of parasites and their application. *Experientia*, **36**, 1369–1371.

Rohde, K. (1989) At least eight types of sense receptors in an endoparasitic flatworm: a counter-trend to sacculinization. *Naturwissenschaften*, **76**, 383–385.

Rohde, K. (1991) Intra- and interspecific interactions in low density populations in resource-rich habitats. *Oikos*, **60**, 91–104.

Rohde, K. (1992) Latitudinal gradients in species diversity: the search for the primary cause. *Oikos*, **65**, 514–527.

Rohde, K. (1993) *Ecology of Marine Parasites*, 2nd edn, CAB International, Wallingford, UK.

Rohde, K. (1994a) The origins of parasitism in the Platyhelminthes. *International Journal for Parasitology*, **24**, 1099–1115.

Rohde, K. (1994b) Niche restriction in parasites: proximate and ultimate causes. *Parasitology*, **109**, S69–S84.

Rohde, K. (1996) Robust phylogenies and adaptive radiations: a critical examination of methods used to identify key innovations. *American Naturalist*, **148**, 481–500.

Rondelaud, D. and Barthe, D. (1987) *Fasciola hepatica* L.: étude de la productivité d'un sporocyste en fonction de la taille de *Lymnaea truncatula* Müller. *Parasitology Research*, **74**, 155–160.

Rosen, R. and Dick, T.A. (1983) Development and infectivity of the procercoid of *Triaenophorus crassus* Forel and mortality of the first intermediate host. *Canadian Journal of Zoology*, **61**, 2120–2128.

Rosen, R. and Dick, T.A. (1984) Growth and migration of plerocercoids of *Triaenophorus crassus* Forel and pathology in experimentally infected whitefish, *Coregonus clupeaformis*. *Canadian Journal of Zoology*, **62**, 203–211.

Rothschild, M. and Clay, T. (1952) *Fleas, Flukes & Cuckoos*, Collins, London.

Rousset, F., Thomas, F., De Meeüs, T. and Renaud, F. (1996) Inference of parasite induced host mortality from distributions of parasite loads. *Ecology*, **77**, 2203–2211.

Rózsa, L. (1992) Endangered parasite species. *International Journal for Parasitology*, **22**, 265–266.

Rózsa, L. (1993) Speciation patterns of ectoparasites and "straggling" lice. *International Journal for Parasitology*, **23**, 859–864.

Rózsa, L. (1997) Adaptive sex ratio manipulation in *Pediculus humanus capitis*: possible interpretations of Buxton's data. *Journal of Parasitology*, in press.

Rózsa, L., Rékási, J. and Reiczigel, J. (1996) Relationship of host coloniality to the population ecology of avian lice (Insecta: Phthiraptera). *Journal of Animal Ecology*, **65**, 242–248.

Sage, R.D., Heyneman, D., Lim, K.-C. and Wilson, A.C. (1986) Wormy mice in a hybrid zone. *Nature*, **324**, 60–63.

Sakanari, J.A. and Moser, M. (1990) Adaptation of an introduced host to an indigenous parasite. *Journal of Parasitology*, **76**, 420–423.

Saladin, K.S. (1979) Behavioral parasitology and perspectives on miracidial host-finding. *Zeitschrift für Parasitenkunde*, **60**, 197–210.

Schad, G.A. and Anderson, R.M. (1985) Predisposition to hookworm infection in humans. *Science*, **228**, 1537–1540.

Schall, J.J. (1989) The sex ratio of *Plasmodium* gametocytes. *Parasitology*, **98**, 343–350.

Schallig, H.D.F.H., Sassen, M.J.M., Hordijk, P.L. and De Jong-Brink, M. (1991) *Trichobilharzia ocellata*: influence of infection on the fecundity of its intermediate snail host *Lymnaea stagnalis* and cercarial induction of the release of schistosomin, a snail neuropeptide antagonizing female gonadotropic hormones. *Parasitology*, **102**, 85–91.

Schluter, D. (1984) A variance test for detecting species associations, with some example applications. *Ecology*, **65**, 998–1005.

Schmidt, G.D. and Roberts, L.S. (1989) *Foundations of Parasitology*, 4th edn, Mosby, St. Louis.

Scott, M.E. (1987) Temporal changes in aggregation: a laboratory study. *Parasitology*, **94**, 583–595.

Scott, M.E. and Anderson, R.M. (1984) The population dynamics of *Gyrodactylus bullatarudis* (Monogenea) within laboratory populations of the fish host *Poecilia reticulata*. *Parasitology*, **89**, 159–194.

Searcy, D.G. and MacInnis, A.J. (1970) Measurements by DNA renaturation of the genetic basis of parasitic reduction. *Evolution*, **24**, 796–806.

Seed, J.R. and Sechelski, J.B. (1996) The individual host, a unique evolutionary island for rapidly dividing parasites: a theoretical approach. *Journal of Parasitology*, **82**, 263–267.

Shaw, D.J. and Dobson, A.P. (1995) Patterns of macroparasite abundance and aggregation in wildlife populations: a quantitative review. *Parasitology*, **111**, S111–S133.

Shine, R. (1989) Ecological causes for the evolution of sexual dimorphism: a review of the evidence. *Quarterly Review of Biology*, **64**, 419–461.

Shoop, W.L. (1988) Trematode transmission patterns. *Journal of Parasitology*, **74**, 46–59.

Shostak, A.W. and Dick, T.A. (1987) Individual variability in reproductive success of *Triaenophorus crassus* Forel (Cestoda: Pseudophyllidea), with comments on use of the Lorenz curve and Gini coefficient. *Canadian Journal of Zoology*, **65**, 2878–2885.

Shostak, A.W. and Dick, T.A. (1989) Variability in timing of egg hatch of *Triaenophorus crassus* Forel (Cestoda: Pseudophyllidea) as a mechanism increasing temporal dispersion of coracidia. *Canadian Journal of Zoology*, **67**, 1462–1470.

Shostak, A.W. and Esch, G.W. (1990) Photocycle-dependent emergence by cercariae of *Halipegus occidualis* from *Helisoma anceps*, with special reference to cercarial emergence patterns as adaptations for transmission. *Journal of Parasitology*, **76**, 790–795.

Shostak, A.W. and Scott, M.E. (1993) Detection of density-dependent growth and fecundity of helminths in natural infections. *Parasitology*, **106**, 527–539.

Shutler, D., Bennett, G.F. and Mullie, A. (1995) Sex proportions of *Haemoproteus* blood parasites and local mate competition. *Proceedings of the National Academy of Sciences of the U.S.A.*, **92**, 6748–6752.

Sibly, R.M. and Calow, P. (1986) *Physiological Ecology of Animals: An Evolutionary Approach*, Blackwell Scientific Publications, Oxford.

Siddall, M.E., Brooks, D.R. and Desser, S.S. (1993) Phylogeny and the reversibility of parasitism. *Evolution*, **47**, 308–313.

Silver, B.B., Dick, T.A. and Welch, H.E. (1980) Concurrent infections of *Hymenolepis diminuta* and *Trichinella spiralis* in the rat intestine. *Journal of Parasitology*, **66**, 786–791.

Simberloff, D. (1990) Free-living communities and alimentary tract helminths: hypotheses and pattern analyses, in *Parasite Communities: Patterns and*

Processes (eds G.W. Esch, A.O. Bush and J.M. Aho), Chapman & Hall, London, pp. 289–319.

Simberloff, D. and Moore, J. (1996) Community ecology of parasites and free-living animals, in *Host-Parasite Evolution: General Principles and Avian Models* (eds D.H. Clayton and J. Moore), Oxford University Press, Oxford, pp. 174–197.

Skorping, A., Read, A.F. and Keymer, A.E. (1991) Life history covariation in intestinal nematodes of mammals. *Oikos*, **60**, 365–372.

Smith, E.N. (1984) Alteration of behavior by parasites: a problem for evolutionists. *Creation Research Society Quarterly*, **21**, 124.

Smith, G. and Grenfell, B.T. (1985) The population biology of *Ostertagia ostertagi*. *Parasitology Today*, **1**, 76–81.

Smith Trail, D.R. (1980) Behavioral interactions between parasites and hosts: host suicide and the evolution of complex life cycles. *American Naturalist*, **116**, 77–91.

Smithers, S.R. and Terry, R.J. (1969) The immunology of schistosomiasis. *Advances in Parasitology*, **7**, 41–93.

Snyder, S.D. and Janovy, J. Jr. (1996) Behavioral basis of second intermediate host specificity among four species of *Haematoloechus* (Digenea: Haematoloechidae). *Journal of Parasitology*, **82**, 94–99.

Sokal, R.R. and Rohlf, F.J. (1981) *Biometry*, 2nd edn, W.H. Freeman & Co., New York.

Sorci, G., Morand, S. and Hugot, J.-P. (1997) Host-parasite coevolution: comparative evidence for covariation of life history traits in primates and oxyurid parasites. *Proceedings of the Royal Society of London B*, **264**, 285–289.

Sousa, W.P. (1983) Host life history and the effect of parasitic castration on growth: a field study of *Cerithidea californica* Haldemann (Gastropoda: Prosobranchia) and its trematode parasites. *Journal of Experimental Marine Biology and Ecology*, **73**, 273–296.

Sousa, W.P. (1992) Interspecific interactions among larval trematode parasites of freshwater and marine snails. *American Zoologist*, **32**, 583–592.

Sousa, W.P. (1993) Interspecific antagonism and species coexistence in a diverse guild of larval trematode parasites. *Ecological Monographs*, **63**, 103–128.

Southwood, T.R.E. (1978) *Ecological Methods, with Particular Reference to the Study of Insect Populations*, John Wiley & Sons, New York.

Southwood, T.R.E. (1988) Tactics, strategies and templets. *Oikos*, **52**, 3–18.

Sprent, J.F.A. (1992) Parasites lost. *International Journal for Parasitology*, **22**, 139–151.

Stankiewicz, M., Jowett, G.H., Roberts, M.G., Heath, D.D., Cowan, P., Clark, J.M., Jowett, J. and Charleston, W.A.G. (1996) Internal and external parasites of possums (*Trichosurus vulpecula*) from forest and farmland, Wanganui, New Zealand. *New Zealand Journal of Zoology*, **23**, 345–353.

Stanley, S.M. (1973) An explanation for Cope's rule. *Evolution*, **27**, 1–26.

Stearns, S.C. (1989) Tradeoffs in life history evolution. *Functional Ecology*, **3**, 259–268.

Stearns, S.C. (1992) *The Evolution of Life Histories*, Oxford University Press, Oxford.

Stepien, C.A. and Brusca, R.C. (1985) Nocturnal attacks on nearshore fishes in southern California by crustacean zooplankton. *Marine Ecology – Progress Series*, **25**, 91–105.

Stien, A., Halvorsen, O. and Leinaas, H.-P. (1996) Density-dependent sex ratio in *Echinomermella matsi* (Nematoda), a parasite of the sea urchin *Strongylocentrotus droebachiensis*. *Parasitology*, **112**, 105–112.

Stock, T.M. and Holmes, J.C. (1987) Host specificity and exchange of intestinal helminths among four species of grebes (Podicipedidae). *Canadian Journal of Zoology*, **65**, 669–676.

Stock, T.M. and Holmes, J.C. (1988) Functional relationships and microhabitat distributions of enteric helminths of grebes (Podicipedidae): the evidence for interactive communities. *Journal of Parasitology*, **74**, 214–227.

Sukhdeo, M.V.K. (1990a) The relationship between intestinal location and fecundity in adult *Trichinella spiralis*. *International Journal for Parasitology*, **21**, 855–858.

Sukhdeo, M.V.K. (1990b) Habitat selection by helminths: a hypothesis. *Parasitology Today*, **6**, 234–237.

Sukhdeo, M.V.K. and Mettrick, D.F. (1987) Parasite behaviour: understanding platyhelminth responses. *Advances in Parasitology*, **26**, 73–144.

Sukhdeo, M.V.K. and Sukhdeo, S.C. (1994) Optimal habitat selection by helminths within the host environment. *Parasitology*, **109**, S41-S55.

Sukhdeo, S.C., Sukhdeo, M.V.K., Black, M.B. and Vrijenhoek, R.C. (1997) The evolution of tissue migration in parasitic nematodes (Nematoda: Strongylida) inferred from a protein-coding mitochondrial gene. *Biological Journal of the Linnean Society*, **61**, 281–298.

Szalai, A.J. and Dick, T.A. (1989) Differences in numbers and inequalities in mass and fecundity during the egg-producing period for *Raphidascaris acus* (Nematoda: Anisakidae). *Parasitology*, **98**, 489–495.

Tanguay, G.V. and Scott, M.E. (1992) Factors generating aggregation of *Heligmosomoides polygyrus* (Nematoda) in laboratory mice. *Parasitology*, **104**, 519–529.

Terry, R.J. (1994) Human immunity to schistosomes: concomitant immunity? *Parasitology Today*, **10**, 377–378.

Théron, A. (1984) Early and late shedding patterns of *Schistosoma mansoni* cercariae: ecological significance in transmission to human and murine hosts. *Journal of Parasitology*, **70**, 652–655.

Théron, A., Gérard, C. and Moné, H. (1992) Early enhanced growth of the digestive gland of *Biomphalaria glabrata* infected with *Schistosoma mansoni*: side effect or parasite manipulation? *Parasitology Research*, **78**, 445–450.

Thomas, F., Mete, K., Helluy, S., Santalla, F., Verneau, O., De Meeüs, T., Cézilly, F. and Renaud, F. (1997) Hitch-hiker parasites or how to benefit from the strategy of another parasite. *Evolution*, **51**, 1316–1318.

Thomas, F., Renaud, F., Rousset, F., Cézilly, F. and De Meeüs, T. (1995) Differential mortality of two closely related host species induced by one parasite. *Proceedings of the Royal Society of London B*, **260**, 349–352.

Thompson, J.N. (1994) *The Coevolutionary Process*, University of Chicago Press, Chicago.

Thompson, R.C.A. and Lymbery, A.J. (1996) Genetic variability in parasites and host-parasite interactions. *Parasitology*, **112**, S7-S22.

Thompson, S.N. and Kavaliers, M. (1994) Physiological bases for parasite-induced alterations of host behaviour. *Parasitology*, **109**, S119-S138.

Thomson, J.D. (1980) Implications of different sorts of evidence for competition. *American Naturalist*, **116**, 719–726.

Tingley, G.A. and Anderson, R.M. (1986) Environmental sex determination and density-dependent population regulation in the entomogenous nematode *Romanomermis culicivorax*. *Parasitology*, **92**, 431–449.

Tinsley, R.C. (1990) Host behaviour and opportunism in parasite life cycles, in *Parasitism and Host Behaviour* (eds C.J. Barnard and J.M. Behnke), Taylor & Francis, London, pp. 158–192.

Toft, C.A. and Karter, A.J. (1990) Parasite-host coevolution. *Trends in Ecology & Evolution*, **5**, 326–329.

Touassem, R. and Théron, A. (1989) *Schistosoma rodhaini*: dynamics and cercarial production for mono- and pluri-miracidial infections of *Biomphalaria glabrata*. *Journal of Helminthology*, **63**, 79–83.

Urdal, K., Tierney, J.F. and Jakobsen, P.J. (1995) The tapeworm *Schistocephalus solidus* alters the activity and response, but not the predation susceptibility of infected copepods. *Journal of Parasitology*, **81**, 330–333.

Uznanski, R.L. and Nickol, B.B. (1982) Site selection, growth and survival of *Leptorhynchoides thecatus* (Acanthocephala) during the prepatent period in *Lepomis cyanellus*. *Journal of Parasitology*, **68**, 686–690.

Vance, S.A. (1996) Morphological and behavioural sex reversal in mermithid-infected mayflies. *Proceedings of the Royal Society of London B*, **263**, 907–912.

Verneau, O., Renaud, F. and Catzeflis, F.M. (1991) DNA reassociation kinetics and genome complexity of a fish (*Psetta maxima*: Teleostei) and its gut parasite (*Bothriocephalus gregarius*: Cestoda). *Comparative Biochemistry and Physiology*, **99B**, 883–886.

Vickery, W.L. and Poulin, R. (1997) Host speciation, parasite extinction and colonization, and the evolution of parasite communities, *Journal of Parasitology*, in press.

Wakelin, D. (1978) Genetic control of susceptibility and resistance to parasitic infection. *Advances in Parasitology*, **16**, 219–308.

Wakelin, D. (1985) Genetic control of immunity to helminth infections. *Parasitology Today*, **1**, 17–23.

Waller, P.J. (1993) Nematophagous fungi: prospective biological control agents of animal parasitic nematodes? *Parasitology Today*, **9**, 429–431.

Walther, B.A., Cotgreave, P., Price, R.D., Gregory, R.D. and Clayton, D.H. (1995) Sampling effort and parasite species richness. *Parasitology Today*, **11**, 306–310.

Wassom, D.L., Dick, T.A., Arnason, N., Strickland, D. and Grundmann, A.W. (1986) Host genetics: a key factor in regulating the distribution of parasites in natural host populations. *Journal of Parasitology*, **72**, 334–337.

Waters, A.P., Higgins, D.G. and McCutchan, T.F. (1991) *Plasmodium falciparum* appears to have arisen as a result of lateral transfer between avian and human hosts. *Proceedings of the National Academy of Sciences of the U.S.A.*, **88**, 3140–3144.

Watters, G.T. (1992) Unionids, fishes, and the species-area curve. *Journal of Biogeography*, **19**, 481–490.

Werren, J.H., Nur, U. and Wu, C.-I. (1988) Selfish genetic elements. *Trends in Ecology & Evolution*, **3**, 297–302.

Wesenburg-Lund, C. (1931) Contributions to the development of the Trematoda Digenea. Part I. The biology of *Leucochloridium paradoxum*. *Mémoires de l'Académie Royale des Sciences et des Lettres de Danemark, Section des Sciences (series 9)*, **4**, 90–142.

Wharton, D.A. (1986) *A Functional Biology of Nematodes*, Croom Helm, London.

Wickler, W. (1976) Evolution-oriented ethology, kin selection, and altruistic parasites. *Zeitschrift für Tierpsychologie*, **42**, 206–214.

Williams, E.H. Jr. and Bunkley-Williams, L. (1994) Four cases of unusual crustacean-fish associations and comments on parasitic processes. *Journal of Aquatic Animal Health*, **6**, 202–208.

Williams, G.C. and Nesse, R.M. (1991) The dawn of Darwinian medicine. *Quarterly Review of Biology*, **66,** 1–22.

Woolaston, R.R. and Baker, R.L. (1996) Prospects of breeding small ruminants for resistance to internal parasites. *International Journal for Parasitology*, **26,** 845–855.

Worthen, W.B. (1996) Community composition and nested-subset analyses: basic descriptors for community ecology. *Oikos*, **76,** 417–426.

Worthen, W.B. and Rohde, K. (1996) Nested subset analyses of colonization-dominated communities: metazoan ectoparasites of marine fishes. *Oikos*, **75,** 471–478.

Wright, D.H. and Reeves, J.H. (1992) On the meaning and measurement of nestedness of species assemblages. *Oecologia*, **92,** 416–428.

Yan, G., Stevens, L. and Schall, J.J. (1994) Behaviorial changes in *Tribolium* beetles infected with a tapeworm: variation in effects between beetle species and among genetic strains. *American Naturalist*, **143,** 830–847.

Yoshikawa, H., Yamada, M., Matsumoto, Y. and Yoshida, Y. (1989) Variations in egg size of *Trichuris trichiura*. *Parasitology Research*, **75,** 649–654.

Zavras, E.T. and Roberts, L.S. (1985) Developmental physiology of cestodes: cyclic nucleotides and the identity of putative crowding factors in *Hymenolepis diminuta*. *Journal of Parasitology*, **71,** 96–105.

Zervos, S. (1988a) Population dynamics of a thelastomatid nematode of cockroaches. *Parasitology*, **96,** 353–368.

Zervos, S. (1988b) Evidence for population self-regulation, reproductive competition and arrhenotoky in a thelastomatid nematode of cockroaches. *Parasitology*, **96,** 369–379.

Index